MW01039671

THE SOVIET SPACE PROGRAM

THE LUNAR MISSION YEARS: 1959–1976

Eugen Reichl

4880 Lower Valley Road • Atglen, PA 19310

Originally published as *Moskaus Mondprogramm* by Motorbuch Verlag, Stuttgart, Germany © 2017 Motorbuch Verlag, Stuttgart, Germany.
www.paul-pietsch-verlage.de
Translated from the German by David Johnston

Library of Congress Control Number: 2017955750

Cover design by Molly Shields
Type set in Avenir LT Std/Univers LT 47 CondensedLt

ISBN: 978-0-7643-5675-9
Printed in China

Published by Schiffer Publishing, Ltd.
4880 Lower Valley Road
Atglen, PA 19310
Phone: (610) 593-1777; Fax: (610) 593-2002
E-mail: Info@schifferbooks.com
Web: www.schifferbooks.com

CONTENTS

FOREWORD

With a haste almost unimaginable by today's standards, amazingly unstructured and always directed only at the immediate goal, from September 1958 onward the Soviet Union sent rocket after rocket into Earth orbit. All too often the launches failed in the first few feet. This book follows in chronological sequence the fifty-nine missions that Moscow sent or intended to send directly to the moon. In addition to Earth orbital flights, these included test missions by the Zond mockup and missions by cosmonauts whose purpose was to prepare the way for the Soviet manned landing on the moon.

In 1974, the Soviets finally abandoned the idea of a manned landing on the moon and for the following decade and a half claimed that they never conducted such a program. The unmanned Soviet lunar flights continued until 1976, by which time they had used up all of the space probes built in the years previous. No more new ones were built. Since Luna 24, the last sample return mission launched in August 1976, no Soviet or Russian spacecraft has flown to the moon.

One must keep in mind that in addition to this tremendous number of unmanned lunar flights and preparatory missions for the manned moon landing, during this time the Soviet Union also sent more than three dozen space probes to Venus and Mars, or at least attempted to; here, too, their efforts were relatively rarely crowned by success.

In the volume you will find a precise chronological sequence of the Soviet lunar missions, from the first experimental launch in 1958 until the complete cessation of flights in 1976. Also, in order to show what was happening in the US space program at the same time, we have placed blue-boxed sidebars throughout the text.

Eugen Reichl

This photo of Earth was taken by the Soviet lunar orbital space probe Zond 7. It took this picture on August 11, 1969. On that day the spacecraft was orbiting the moon at a distance of about 1,240 miles.

The goal of the Zond program had been to carry out a manned orbital flight of the moon before the US could; however, the Zond spacecraft were initially not safe enough to entrust humans to them. By the time they finally were, the Americans had long since begun sending manned vehicles to the moon and had even landed on it. The Soviets therefore saw no point in carrying out manned Zond flights. In the perspective of that time, this only confirmed their failure.

INTRODUCTION

In 1952, under the direction of Sergey Korolev, the Soviet Union began development of the world's first intercontinental rocket, the R-7. This was an extremely ambitious undertaking. The first test flights by the R-5 medium-range rocket, also a product of Korolev's OKB-1, were taking place at this time. In its day, the R-5 represented the apex of rocket technology and was capable of carrying an 80-kiloton nuclear warhead. It was a single-stage rocket, weighed about 33 tons at launch, and was capable of transporting its 2,200-pound nuclear payload a distance of 750 miles.

This 80-kiloton warhead also initially defined the payload requirement for the R-7. Its range had been defined by a government

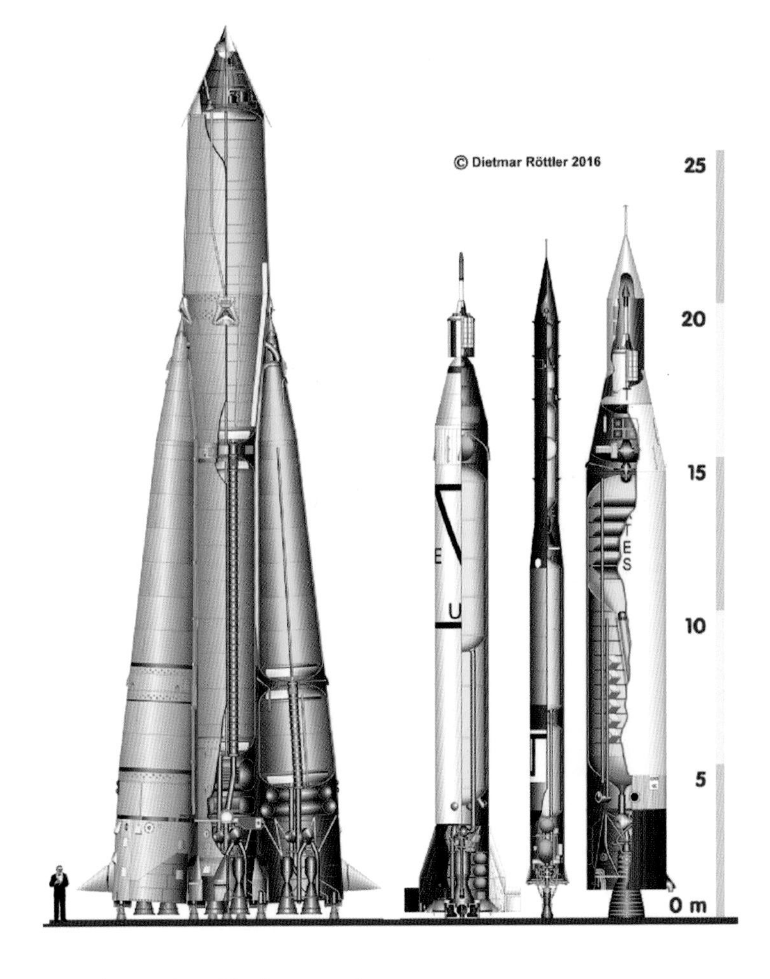

Comparison of several satellite boosters from the year 1958. From left, *R-7 (Soviet Union), Jupiter C (US), Vanguard (US), and Juno 2 (US).*

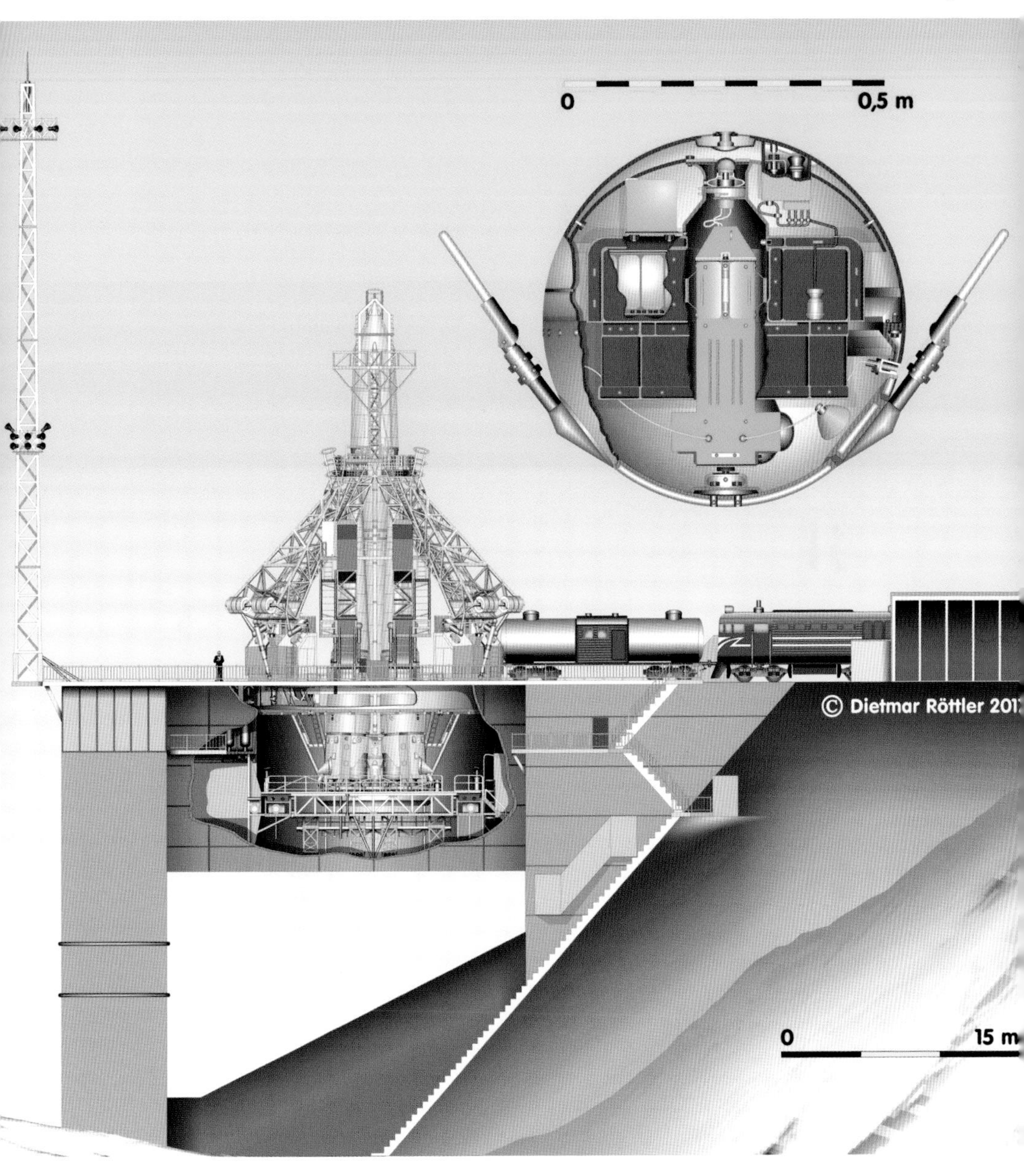

Diagram of the Sputnik R-7 8K72 booster rocket with Sputnik 1 on launchpad 1/5 at Baikonur on the classic Tyulpan launch platform. The first Luna probes were also launched from there. Above right, a cutaway view of Sputnik 1.

decree in 1953: 5,000 miles. This was far enough to reach the United States from central Russia over the North Pole. The R-5's payload weight of 2,200 pounds rose to more than 6,600 pounds in the R-7, since its considerably higher flight speeds made it necessary to house the nuclear warhead in a heat-resistant reentry body to prevent it from burning up during reentry into Earth's atmosphere. To achieve these performance figures, Korolev determined that the R-7 would have to have two stages. He calculated that its launch weight would be 187 tons, six times as heavy as the R-5.

Then, however, something happened that would determine the history of the USSR in the first years of space flight. On August 12, 1953, the Soviet Union successfully detonated its first thermonuclear device at the nuclear-weapons test site at Semipalatinsk. One of the test directors was Vyacheslav Malyshev, member of the Presidium of the Central Committee of the Communist Party of the Soviet Union and, since 1953, head of the Ministry of Medium Mechanical Engineering. In this function he was one of the main decision makers not only for development of the hydrogen bomb, but also for Soviet rocket construction.

He thus knew about the secret project on which Korolev was working, and he knew the basic specification of the R-7. When it was proved on that August day at Semipalatinsk that Soviet scientists had mastered the principle of the hydrogen bomb, he confronted them with the demand that they reduce the size and weight of the weapon so that it could be carried by Korolev's new rocket. But hard as they tried, the weight and dimensions remained clearly greater than those of a fission bomb. They could not reduce the weight below 3.3 tons.

In May 1954, Malyshev went to Korolev and confronted him with the new parameters. The 3.3 tons for the bomb alone meant six tons for the entire warhead and heat shield, the activation mechanism, and the adapter elements for the rocket. Malyshev demanded that the R-7's range remain 5,000 miles.

Korolev worked out the figures, and the result was shocking. The design, which was already complete, had to be totally revised. The rocket would have to weigh an additional 110 tons to carry this massive payload over such a range. Instead of six times as heavy as the previously largest rocket in the world, it would now have to be designed to be nine times as heavy.

In his autobiography, Boris Chertok, one of Korolev's companions in the OKB-1, wrote that Korolev and his designers broke out into fits of rage when they had to accept the new demands, which were contained in government resolutions on May 20 and June 28, 1954. It appeared that what had been a seemingly hopeless task had become a completely impossible one.

Korolev's massive criticisms of the new specification soon fell silent. He quickly realized the tremendous possibilities that lay in this almost impossible commission. He would create not only a large intercontinental missile, but also a powerful vehicle for the carriage of spacecraft. With its enormous power, many operational variants of the R-7 would dominate the first decade of the space age, and even sixty years later it would remain in use in its ultimate incarnation.

FROM THE 8K71 TO THE 8K72 LUNA

On June 30, 1956, in the midst of development of the R-7, the government issued a resolution for the production of a satellite with the designation Object D, with a weight between 2,200 and 3,000 pounds and a scientific payload of 440 to 660 pounds. This spacecraft, the later Sputnik 3, was supposed to be placed in orbit in 1957 or 1958. It was planned as Moscow's contribution to the International Geophysical Year. Its task was not only to preempt the satellites already announced by the Americans, but to destroy them with its sheer mass and technical complexity.

In the last days of 1956, barely half a year before the rocket's first flight, the engine began test stand trials. These showed that Glushko's RD-107 and RD-108 engines did not deliver the performances he had promised. Their specific impulse was only 304 seconds instead of the promised 310 seconds. This was of no great importance to the R-7's function as an intercontinental ballistic missile, but it seriously affected its use as a satellite launch vehicle. Instead of the previously estimated 3,306 pounds for an elliptical Earth orbit with an perigee of 137 miles and an apogee of 621 miles, now only about 175 to 220 pounds remained, and this only if the majority to the telemetry equipment was removed and the burn termination procedure was "tuned." This was sufficient for Sputnik 1 and Sputnik 2, which—as is scarcely known today—were not separated from the booster rocket. Sputnik 2 remained firmly attached to the center stage. Dispensing with the separation elements was simply a weight-saving measure, and they fell back on the rocket's Tral telemetry system for radio transmission.

In any case, Glushko promised that the next production batch, beginning with production no. "13," would have more-powerful engines. This was, however, too late to place Object D into orbit in autumn 1957. It turned out, however, that the satellite was too complex for the launch target to have been met. The first attempt to place Sputnik 3—or Object D—into orbit took place on April 27, 1958. The rocket broke up 338 seconds after launch, and its wreckage and that of the valuable satellite rained down on the steppes of Kazakhstan. Incidentally, the variants of the R-7 procured specifically for Object D bore the designation 8A91. They were used only for this purpose and afterward never flew again.

A few words about the rockets' nomenclature are appropriate here. The designation R-7 appears nowhere in the technical documentation, nowhere on a design drawing, and nowhere in the internal correspondence. It also rarely appears in the countless secret documents. Only the "article number" is found there, and for the first test version of the rocket this was 8K71. The name "R-7" does, however, appear in almost all governmental decrees, resolutions, and other official correspondence. Scarcely anyone there described it by using the article number.

It was also common practice to use only the last two digits in normal interactions between the engineers and technicians. They therefore spoke about the number 71, the 91, the 72, or the 74. And the name *Semyorka* (good old seven) was generally used in verbal communication when speaking of the R-7 in general and not a specific version. This was a follow-on to the practice adopted for the R-5 medium-range

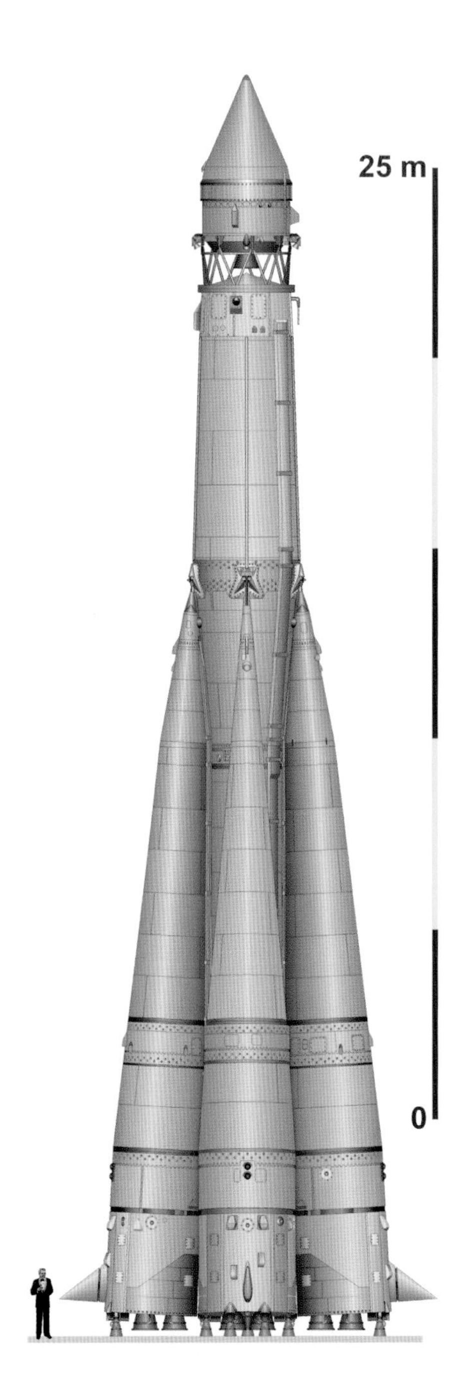

The 8K72 Luna booster rocket.

rocket, which had been referred to informally as the *Pyatyorka*, or "good old five."

Nowadays these names are common among spaceflight historians, but at the time the constantly changing names must have been a headache for the CIA. This was complicated by the fact that even minor modifications resulted in new article designations. The R-7s for Sputnik 1 and 2, which were changed only slightly, were covered by the article number 8K71PS, and even the rocket for Sputnik 3, as we have seen, received its own designation—8A91—even though it was modified exclusively for the launch of this satellite and in these variants was never used again.

The first major redesign resulted in the 8K72, which first flew on September 23, 1958. It had the improved engines and many other modifications, which allowed the R-7's range to be increased to more than 7,500 miles. This was not, however, the case on its first mission and also not on the three that followed, all of which failed due to excessive vibration. The first Luna probes were ultimately launched by a version of this rocket with an additional third stage. And finally, there was the 8K74, sometimes also called the R-7A. It was the production version of the operational ICBM.

At the beginning of 1959, the future of the R-7 was hanging in the balance. Its reliability was simply too poor, and in the Defense Ministry serious consideration was being given to halting the R-7 program. Michael Jangel's R-16 was nearing its first flight, and with it an alternative for an intercontinental missile would be in hand. Between May 24, 1958, and January 2, 1959, there were no fewer than seven failed launches in a row by the R-7, and only a feat of accounting saved the program from being canceled in those days. In determining the failure rate, the Defense Ministry had been convinced to deduct

the first three Luna launches from the failures, even though their first and second stages were identical to those of the military 8K72, and during these flights it had been the same first and second stages that had failed. The rocket for the fourth Luna mission was, however, counted among the successes (the third stage's control system had failed here), with the justification that the first two stages had functioned and were after all identical to the military version.

These considerations were just one brief episode, however. As quickly as it became clear that the R-7 would have no future as a military rocket because of its size, its complexity, and the fuel it used, it just as quickly became clear that its future would lie in its use as a booster rocket for space vehicles.

1958

To mark the forty-first anniversary of the October Revolution, and the first anniversary of the launching of the first Sputnik, Nikita Khrushchev, first secretary of the Communist Party, demanded a new major space event, something with which to once again demonstrate to the West the superiority of the Soviet system. He therefore issued a directive on May 20, 1958, calling first for a probe to be sent to the surface of the moon, followed by a photographic mission to the far side of the moon. This directive was the initiator of the first series of lunar flights by the USSR.

The Soviet Union subsequently undertook nine launches of Earth satellites between September 23, 1958, and April 4, 1960. Two missions accomplished their assigned tasks, and a third was a rather involuntary partial success. The remaining six flights were failures. What looks like a miserable statistic from the vantage point of the present day is, however, a relatively good success rate when one bears in mind that of the subsequent flights of the following series, a total of thirteen by January 1966, only two of which were completely successful.

One major problem confronting Soviet designers was the fact that the two-stage R-7

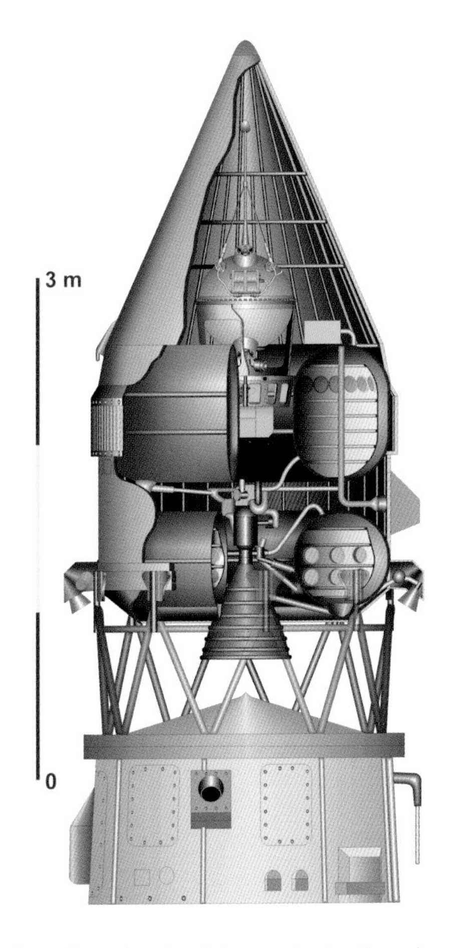

Luna 1 on the tip of the newly developed Block E upper stage.

was incapable of reaching the moon. To achieve the necessary velocity of at least 24,382 miles per hour, the rocket needed an additional propulsion unit. The main difficulty in achieving this was that Khruschev's plan left just an extremely small amount of development time.

The new third stage was designated Block E, corresponding to the sixth letter of the Cyrillic alphabet. Block A was the R-7's central stage. Blocks B, V, G, and D—which corresponded to the next letters of the Cyrillic alphabet—were the four boosters. Thus the next available letter was "*Ye,*" or, in Western transcription, E. In the Soviet point of view the boosters formed the first stage, Block A was the second stage (even though it was ignited simultaneously with the boosters), and Block E was the third stage.

The engine for the Block E was not provided by Valentin Glushko of OKB-456, even though he had a virtual monopoly for rocket propulsion units in the Soviet Union. Cooperation with Glushko was out of the question for Korolev. Over the years a growing hostility had developed between the two men, and moreover Glushko favored engines with storable fuels. Their disadvantage was that they worked with toxic components and had a lower specific impulse power than power plants that were powered by kerosene and oxygen. And so, Korolev turned to Semyon Kosberg of OKB-154 for help.

In principle, OKB-1 entrusted development of the Block E power plant to itself, since under Vasily Mishkin it had already designed the R-7's attitude control and roll engine. This rocket engine now formed the basis of the Block E engine, which was given the designation RD-105. As desired by Korolev, it was powered by kerosene and liquid oxygen. The problems, however, were the turbopumps, with which OKB-1 had only limited experience.

Kosberg had little previous experience with rocket engines. He had worked for the aviation industry and had mainly developed and produced components for jet engines. And so the development work on the Block E propulsion system was divided. Kosberg developed the turbopumps, the gas generator, and the connecting structures for the main components, while OKB-1 concerned itself with the thrust chamber and the nozzle. For the first time in spaceflight, this was designed exclusively for operation in a vacuum.

The collaboration between Kosberg and Korolev did not begin in early February 1958. Once again, the impossibly short periods in which developments took place in those days (by today's standards) are fascinating; however, only in very few instances were they really fully developed.

The engine developed just under 11,240 pounds of thrust and was designed to operate for six and a half minutes. The biggest problem faced by the developers was the fact that it had to be capable of reliable ignition in space. To circumvent the associated difficulties, the R-7 had been designed with parallel stages. Thus, on the ground it was possible to ignite all the engines simultaneously and then launch the rocket only if they were running stably. If this was not the case, then the launch could be aborted. This was not possible with the

The launch of Luna 2 on January 2, 1959.

Display model of the Block E stage with Luna 1.

engine of the 8K72 Luna, however. It could not be ignited until the second stage burned out.

The developers' answer to this problem was the introduction of "hot" stage separation. It remains in use today by the Russian Soyuz (which is a direct successor to the R-7) and Proton booster rockets. The Ukrainian Zenit rocket, whose origins also date back to the Soviet period, also functions in this way. The third stage ignites while the second propulsion unit is still running, and the exhaust gases are diverted outward through a lattice structure. The advantage of this method is that it avoids complex pre-acceleration engines, which would otherwise be necessary to keep the fuel at the bottom of the tank.

The method of immediate sequential ignition of the third stage, with a direct flight from the launchpad to the moon, had serious disadvantages, however. For one, it did not exploit the rocket's maximum payload capacity, and for another, achieving the precision required for achieving a lunar trajectory was difficult. Only a few years later, it was standard practice both in the Soviet Union and the US to initially head for a parking trajectory in Earth orbit instead of carrying out the translunar acceleration maneuver directly. The more precise method with the parking trajectory did, however, require that the third-stage engine first be shut down and then reactivated when the "injection window" was reached. However, the Soviets did not yet have sufficient mastery of restarting the engine in conditions of weightlessness.

The matter of injection precision was difficult, since at the time of third-stage separation the R-7 booster could not exceed a certain "angle of heel," because otherwise the thrusters of the upper stage could no longer compensate for deviations in attitude. If this problem was circuited, then the stage's thrusters had to keep the stage precisely on course. An accelerometer onboard measured the cumulative speed change and switched off the engine at the calculated time. This direct-launch method also required the time of launch to be accurate and the timing of the burnout even more so. This was no simple exercise in 1958. A deviation of ten seconds during launch meant that the target point on the moon would be missed by 155 miles. A deviation of just 3 feet per second during release of the Luna stage also meant a deviation of 155 miles. And then, should the actual position in space deviate by even 1 angular minute from the calculated value during the more than six-minute burn time by the upper stage, the result would be a deviation of 120 miles. One must keep one thing in mind: these first primitive space probes could not make course changes en route to the moon. They had to be given the necessary precision from the beginning.

If all these targets were achieved, then it could be left to celestial mechanics to do the rest, taking the probe to the moon in about thirty-four hours.

August 17: A Thor-Able carrying the first space probe of the US Air Force's Pioneer Program exploded seventy-seven seconds after liftoff at an altitude of 10 miles. The spacecraft was actually supposed to be called Pioneer 1, but after the failed launch it was renamed Pioneer 0.

Model of Luna 1.

The first test flight by the 8K72 version, the very same variant that was to put the Luna on its path to the moon, took place on July 10, 1958. It was a complete failure. The rocket disintegrated without warning eighty-eight seconds after leaving the launchpad. The cluster of four boosters and the central stage simply broke apart. No definite cause for the failure was found; however, telemetry showed that high-frequency vibrations had developed in one of Block D's combustion chambers immediately before breakup. The investigators could not be sure, however, if this could have led to the observed structural failure in the booster group. There was speculation that it might have been caused by production or assembly errors, but there was nothing in the telemetry data or documentation to support this theory.

The first attempt to send an E-1 unit to the moon took place on September 23, 1958. For the first two stages of the R-7 it was also the second test flight of the 8K72 version. Ninety-two seconds after liftoff, this rocket also disintegrated in the same way as the first 8K72 had two and a half months earlier. The similarity of the two accidents was striking. This time they had not just lost the rocket, but the first Soviet lunar probe as well. The engineers faced a mystery. Analyses revealed no production error, no design error, no shortcomings in the execution of tests, and no sloppiness in assembly. Both failed launches followed a pattern that had never been encountered before. Suspicions grew that perhaps something fundamental had been overlooked in the rocket's design, or that saboteurs might possibly be at work. The designers would almost have preferred this answer, since if it turned out to be a serious technical problem, and then there was no possibility that it might be solved by the next launch.

Mission Data, E-1 No. 1	
Mission designation	—
Date	September 23, 1958
Spacecraft	E-1 no. 1
Booster rocket	8K72 (B1-3) Luna
Spacecraft weight	423 pounds
Planned mission objective	Impact on the moon
Mission results	Mission failed. Booster rocket exploded ninety-two seconds after leaving the launchpad, due to massive pogo vibrations, which led to structural failure

The next launch, and also the last opportunity to give the state and party leadership a "gift" for the anniversary, was scheduled for October 12. For celestial-mechanics reasons it had to be this date precisely and no other. Thus they were forced to launch. The military saw it rather differently, since it did not want to undertake another launch until the causes of the previous failures had been discovered and eliminated. Because the 8K72 was ultimately the precursor to the 8K74, the production version of the R-7 in its role as intercontinental ballistic missile, Korolev and his associates found themselves in a quandary.

When the engineers closely analyzed the sensor data from the crashed B1-3 unit, they were struck by one feature that they had overlooked in their first review, because it seemed to be within normal parameters. Only when they combined it with data from other sensors from other areas of the rocket was an unusual pattern revealed. This biased the conclusions in a certain direction. And the OKB-1 engineers loathed this direction.

The data revealed an unusual behavior by the pressure sensors in the booster block's combustion chambers. They showed that the combustion of the fuel produced a pulse mode

Model of Luna 2.

between 9 and 13 Hertz (Hz), which led to regular pressure variations of plus or minus 4.5 atmospheres in the combustion chambers. But how could that happen? These variations were well within tolerance limits. It was for this reason that the telemetry evaluators had not reported any anomalies.

The mystery was finally solved by much detailed analytical work. The pressure variations in the combustion chambers were transmitted back into the system of pipes and fuel tanks. In doing so they interacted with the natural frequency of the rocket structure in the longitudinal axis. The two amplitudes continued to build up, and all the rocket's fuel pipes, tanks, and structures became caught up in a steadily increasing vicious circle of interlinked resonances, ultimately causing the rocket to break up.

These characteristic longitudinal vibrations in large booster rockets remain the terror of spaceflight engineers to this day. They are referred to as the pogo effect. The Americans first encountered the phenomenon during flight trials of the Thor medium-range missile. They were also not completely unknown to the Soviets; significantly higher-resonance vibrations had appeared during testing of the smaller rockets leading up to the R-7 without, however, causing any damage. The R-7 was much larger and the frequency was lower, but the vibrations were much more powerful than those encountered with the smaller rockets.

The logical decision would have been to immediately stop flight testing and study the problem in detail, but the Soviet spaceflight managers and responsible political figures acted like gamblers. The risks were great, but so were the stakes. The prize was to be the first to send a probe to the surface of the moon. For the designers it was an opportunity to increase the fame of the Soviet Union and their own as well. No one wanted to hold up the flight. No one had any desire to go to the leader of the party and the state and tell him that unfortunately his lovely lunar mission on the forty-first anniversary of the party would come to nothing. And so the decision was made to take the chance of an experimental launch in October.

The series number B1-4 incorporated several hastily devised modifications to prevent the rocket from breaking up again. It was decided to reduce thrust after the eight-fifth second of flight to minimize structural loads. The fuel delivery system was tuned a little in the hope that it would react less sensitively to the vibrations than before. Reinforcements and dampers were added in different places,

in the hope that the structure would be able to withstand greater loads. Other measures served to take the fuel pipes and delivery system out of the resonance zone. The OKB-1 engineers presented these measures to the state committee, which had to authorize the next flight. It agreed with some reluctance and approved the launch.

October 11 and 12: A Thor-Able launched the Pioneer 1 probe. It was planned to place it in lunar orbit. Because a burnout occurred ten seconds too early, Pioneer 1's third stage achieved only a ballistic flight path with an apogee of 70,214 miles, and forty-three seconds after launch it reentered Earth's atmosphere.

And so on October 12, 1958, the R-7 8K72 with the series number B1-4 was launched. This time it endured for 104 seconds, and then it too exploded. Onboard was the second Luna E-1 probe, which went up in flames with the rocket. Khruschev's gift for the forty-first anniversary of the October Revolution was destroyed. Only then did an investigation committee begin analyzing the full extent of the problem. The next launch window for the moon, which opened on November 8, was allowed to pass by, even though the Soviet scientists knew that the Americans were preparing a moon launch for that day.

The investigations now began on a broad and analytical basis. The specialists worked day and night, since time was pressing if the launch window after the next was to be reached. Finally, the starting point of the pressure variations was discovered. It was a valve at the entrance to the oxidizer turbopump. The oscillation began at that point, was reinforced in the combustion chamber area,

Mission Data, E-1 No. 2	
Mission designation	—
Date	October 12, 1958
Spacecraft	E-1 no. 2
Booster rocket	8K72 (B1-4) Luna
Spacecraft weight	423 pounds
Planned mission objective	Impact on the moon
Mission results	Mission failed. Booster rocket exploded 104 seconds after leaving the launchpad, due to massive vibrations. Same problem as B1-3

and then spread through the entire rocket, where it entered into resonance with the booster's structure.

The solution to the problem was ultimately a hydraulic damper, which was installed in the oxidizer line at the pump entrance. The system was complex and had to fit into the fuel line system, which was no simple matter. The system was tested on the test bench, revised several times, and finally released for series production of the R-7.

November 8: Launch of Pioneer 2. The third stage of the Thor-Able booster did not ignite, and therefore the probe reached an apogee of just 963 miles, after which it fell back to Earth. The entire mission lasted just forty-five minutes before the probe burned up over Africa.

On December 4, 1958, the R-7 8K72 with the series number B1-5 was launched. All the measures to suppress the pogo effect worked, the four booster blocks functioned normally, and the stage separation took place when it should have. But this time another fault appeared. A crack developed in a hydrogen-peroxide tank in Block A of the R-7, the one

Mission Data, E-1 No. 3	
Mission designation	—
Date	December 4, 1958
Spacecraft	E-1 no. 3
Booster rocket	8K72 (B1-5) Luna
Spacecraft weight	344 pounds
Planned mission objective	Impact on the moon
Mission results	Mission failed. At 245 seconds after leaving the launchpad, the booster rocket's central stage lost thrust because of the failure of a hydrogen peroxide pump

that powered the central-stage turbopump. After 245 seconds, all the hydrogen-peroxide was gone, and the engine lost power and finally stopped completely. As a result, the third Luna probe was also lost.

December 6–7: A Juno 2 rocket blasted off carrying the Pioneer 3 probe, which weighed barely 13 pounds. It was supposed to make a flypast of the moon and then enter a heliocentric orbit. A deficient performance by the booster's first stage (burnout took place four seconds too soon) and a 3-degree error in the alignment of the flight path inclination angle resulted in the probe getting only 63,380 miles from Earth. After just thirty-four seconds it reentered the atmosphere and burned up over Africa.

1959: FIRST HALF OF THE YEAR

The idea for the future lunar probes essentially came from Mstislav Keldysh, who in 1961 became president of the Academy of Science of the Soviet Union. They were given designations that were oriented toward the names of the R-7 Luna upper stages, and all began with an "E." This nomenclature continued until the end of the Soviet unmanned lunar program. The E-1 probes were simple devices designed only to impact on the moon. The E-2 was a flyby probe that was supposed to photograph the far side of the moon. From today's view the E-3 was the most exotic of all. Its goal was to transport an atomic bomb to the moon and detonate it on the surface. E-4 stood for a circumlunar mission, which was supposed to take high-resolution photographs of the far side of the moon. E-5s were lunar orbital probes, and finally the E-6 series was the crown jewel: a soft landing on the moon. The meaning of this nomenclature changed a little in the later 1970s because it then designated the series rather than the mission. The last series, which was in service until the end of Soviet lunar exploration in 1976, was the E-8-5 series. In 1959, it was calculated that an E-6 probe would be able to make a soft landing on the moon by about 1963.

For the scientists, however, the most fascinating project was the plan involving the atomic bomb. The idea behind it was to produce such a tremendous flash of light on the moon that all observatories on Earth would be able to detect it. This would demonstrate clearly and concisely to all that the Soviet Union had sent an impact probe to the moon, and it would also be a demonstration of military power.

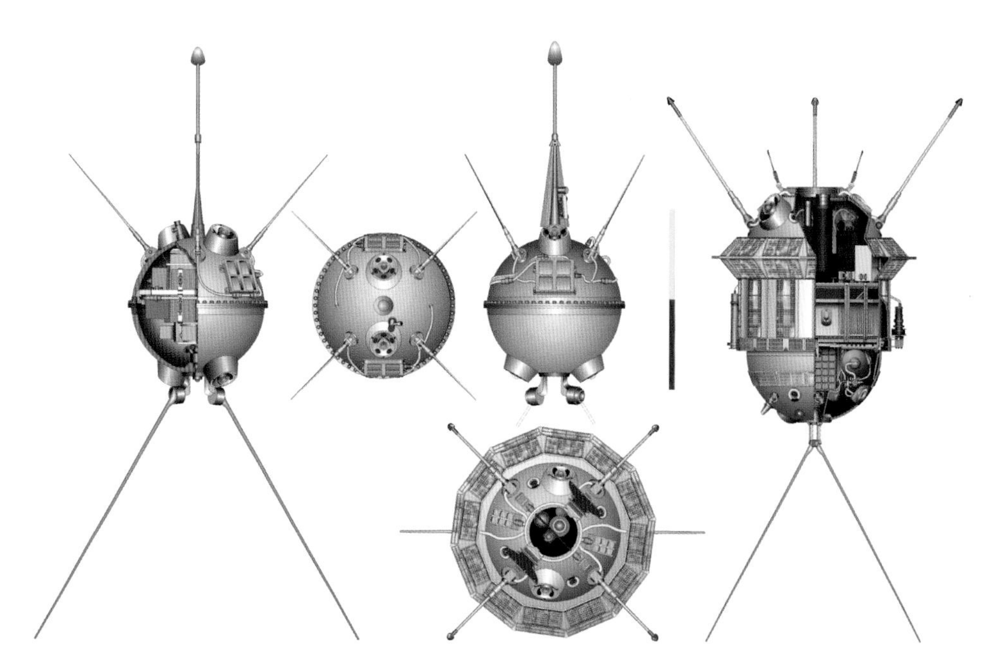

Illustration showing Luna 1 (left)*, Luna 2 from two viewing angles* (both center images)*, and Luna 3 from two viewing angles* (right and bottom)*.*

The OKB-1 engineers immediately set to work and soon created a mockup that resembled a sea mine. It was spherical and from this sphere projected short, rod-shaped detonator pins, which were intended to function as impact detonators. They were located around the sphere to ensure that at least one set off the nuclear explosion, no matter in which attitude the bomb struck the moon.

The initial enthusiasm for this idea soon gave way to a certain disillusionment when the plan was thought through more precisely. Keldysh in particular was opposed to the idea. He observed that he had not the slightest desire to give the world's scientific community the news that the Soviet Union planned to set off a nuclear device on the moon, since that would probably be met with incomprehension. And despite the nuclear explosion, it would have to be announced in advance to ensure that the astronomers aimed their instruments at the moon in time.

Korolev himself was undecided. Instead he was plagued by thoughts of what would happen if the launch failed and the device came down somewhere in Soviet territory. It was less the thought that the device might possibly explode unintentionally. This possibility was unlikely, however. But he did not like the idea that a Soviet atomic bomb might fall somewhere in the Americans' front yard.

Finally the idea by the nuclear experts was itself buried. After everything had been thought through and the brightness of the flash in the vacuum of space at a distance of 240,000 miles had been determined, there

Plaque on Luna 2, in the shape of a moon.

was disillusionment. They feared that photographic equipment on Earth might not even be able to detect the explosion. And so the idea was finally shelved. The number E-3 was subsequently assigned to the planned circumlunar photo mission.

From a technical standpoint, the fourth flight of an 8K72 Luna on January 4, 1959, was largely successful. All three stages functioned perfectly, and had there not been an error in the flight control system, the Soviets would have reached the surface of the moon in these January days of 1959. But in fact the mission was a failure.

The problem with the direct insertion of the first Lunas into the lunar transfer trajectory was the fact that the burnout times for each stage had to be maintained precisely. The engineers did not trust the automatic system onboard the rocket and decided to send radio signals to the Block A central unit and then to the Block E stage, so as to be able to achieve burnout more precisely. Unfortunately, in this case it was the radio station that sent the

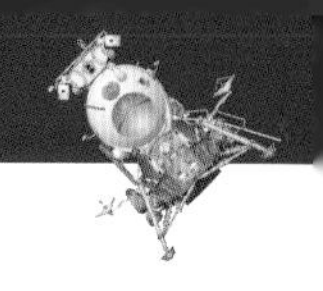

burnout signal too late. And so the Block E stage with the lunar probe missed its target by 3,600 miles, almost one and a half times the diameter of the moon. The stage and the probe entered a heliocentric orbit, and this became "the first artificial planet in the solar system," as it was expressed at the time.

Instead of the anticipated chastisement for having bungled the mission after a successful launch, the engineers were showered with praise and congratulations. The Soviet propaganda system did an excellent job, convincing the world that a flypast of the moon and entry into a heliocentric orbit had been precisely the objective of the mission and had been achieved 100 percent. On January 5, the Central Committee of the Communist Party and the Council of Ministers of the USSR released a message with this wording: "Glory and honor to the workers of Soviet science and technology, who are taking new paths in the discovery of the secrets of nature and are combining its powers for the good of mankind."

In any case, this flight was a good test for later missions. For the first time, the new third stage could be tested over its entire performance spectrum. For the first time, the radio communications system could be tested to a range of 360,000 miles from Earth. The equipment onboard had functioned well, and during a press conference on January 12, the Soviet scientists were able to share their findings with the world.

The greatest sensation for the scientists at that time was the discovery that the moon had its own magnetic field. The press, on the other hand, was more interested in the sodium gas cloud released by the probe at 70,200 miles from Earth. This cloud had a very special purpose. It was easily observed from Earth, and it moved toward the moon at the same speed as the probe and the third stage. It served as proof that the Soviets had in fact reached the moon or, more correctly in this case, had flown past it.

Sixty-two hours after launch, the batteries onboard the probe began giving up their ghost. This also satisfied the engineers, who had designed the batteries for a minimum life of forty hours.

Mission Data, Luna 1	
Mission designation	Luna 1 (also Lunik 1)
Date	January 2, 1959, 17:21 CET
Spacecraft	E-1 no. 4
Booster rocket	8K72 (B1-6) Luna
Spacecraft weight	796 pounds
Planned mission objective	Impact on the moon
Mission results	Flypast of the moon at a distance of approximately 3,600 miles
Last contact	January 5, 1959, at about 10:00 Moscow time at a distance of 370,958 miles from Earth
Current location	Heliocentric orbit (perihelion: 0.98 AE; aphelion: 1.32 AE)

March 3: Pioneer 4 was launched on a lunar flypast trajectory and placed into a heliocentric orbit. However, because of an "overperformance" by the booster's second stage, Pioneer 4 missed the desired flypast distance (less than 18,000 miles) and missed the moon by 36,660 miles. Pioneer 4 was the first American space probe to reach interplanetary space. The probe continued to transmit data from a distance of 409,000 miles, but contact was lost after eighty-three hours.

After Luna 1 there was a short break in lunar missions. Until June, OKB-1 dedicated all its energies toward further development flights by the R-7. Six military test missions were flown from Zemyorka in that period. None of them were completely successful. Eliminating these problems was also very important for the coming flights to the moon and the planets.

The next attempt to send an impact probe to the moon took place on June 18, 1959. The launch, by the 8K72 (production unit I1-7), was a total failure. A leak caused the attitude control gyro in Block A to lose its fluid, the rocket went out of control, and the mission had to be terminated by shutting down the power plant.

Up until that point, the R-7's ratio of failures to successes was miserable, to put it mildly. Of the twenty-four missions since May 15, 1957, only ten had been successful. It was not until after the failed flight on June 18 that the R-7's statistics improved significantly. This was followed by a series of eleven consecutive launches of 8K71 and 8K72 versions, all of which were successes. These included the flight of Luna 2, the fourth R-7 mission after the failed June flight, and of Luna 3, the sixth subsequent flight. All the other missions were test missions for the ICBM version of the rocket.

Mission Data, E-1A No. 1	
Mission designation	—
Date	June 18, 1959
Spacecraft	E-1A no. 1
Booster rocket	8K72 (I1-7) Luna
Spacecraft weight	853 pounds
Planned mission objective	Impact on the moon
Mission results	Mission unsuccessful. The booster rocket's flight control system failed. Flight control shut down the engine 153 seconds after liftoff

1959: SECOND HALF OF THE YEAR

The next moon shot after three successful ICBM tests of the R-7 was scheduled for September 6 at 0349 in the morning. The mission launch window was only ten seconds long. If it was missed, then the mission would have to be postponed for at least twenty-four hours. And so it happened: an electrical circuit automatically deactivated, and after hours of searching it was discovered that someone had wired a plug incorrectly. The reason for this was that the correct connection was not shown in the documentation diagrams. Discovering the error, rewiring and testing the fix took more than a day. And so, the next launch attempt was scheduled for September 8 at 0541 local time.

On that day, everything initially went according to plan. The rocket was again fueled with liquid oxygen (the kerosene remained in the tanks). Finally came the point in the countdown when all the tanks were brought to operating pressure with compressed nitrogen. This succeeded in the boosters and Block E, but not in the central Block A. There, tank pressure refused to move past the 60 percent mark. Time was running out. Finally, it was decided to drain the pressurized gas and try again. But this and a third attempt also failed. The launch had to be called off and was rescheduled for September 9.

This time the liquid oxygen remained in the tanks. The oxygen lost through evaporation had to be constantly replaced during the night. Because of the cold oxygen, actuators and lines had to be warmed regularly. For reasons

of celestial mechanics, the launch that day was supposed to take place at 0639 hours and 50 seconds Moscow time. This time, good luck appeared to be with the Soviet spaceflight engineers. Ignition took place, and in typical R-7 fashion, all the engines were already running on the "preliminary step," the condition in which the turbopumps ran up and the engines built up their thrust. But then the signal for the "main step" did not come through. Once again it was a circuit, which went into reset mode, and the engines died.

Now all hell broke loose on the launchpad. At that time, rarely had there been a rocket whose engines shut down on the pad after they had been run up. All emergency procedures were initiated, fire trucks moved in and positioned themselves, and after a while the rocket was secured. Korolev looked at the situation and then decided on something that was possible only during the Cold War. He simply had the rocket defueled and readied the next one, which was waiting in the hangar.

Just four days later, at 09:39:26 Moscow time, the overall sixth attempt by the Soviet Union to send a probe on a path to the moon's surface succeeded. The deviation from the planned launch window was just one second.

The first task for Korolev, Chertok, Keldysh, and the other responsible figures after the successful launch was to draft a report for the Soviet news agency TASS. Afterward, Mstislav Keldysh had to obtain permission from the state committee to contact Professor Bernard Lovell of the Jodrell Bank Observatory in London. He had to be informed about Soviet intentions with this space probe. Contact with Lovell was of great importance, since at that time the Jodrell Bank Observatory had the world's largest parabolic antenna.

Korolev initially opposed the idea of Keldysh making contact. What if they missed the moon again? This time, no one would believe that the Soviet Union would intentionally plan to send another probe around the sun that would again come very close to the moon. Keldysh nevertheless won out. He gave the Soviet Academy of Science the task of immediately contacting Lovell to transmit to him the calculated impact point on the moon and Luna 2's trajectory parameters so that he would be able to detect the weak signals from the probe against the background radiation of space.

The entire action was based on fear on the part of the Soviet engineers and scientists that possibly no one on Earth might notice the event and in the end would not believe them. Or that while the Americans might notice (which in fact they did), they might possibly tell no one. In fact,

Display model of Luna 3.

ultimately it turned out that the Americans also contacted Lovell to verify their own observations. And it was not just the military authorities that contacted Lovell, but NASA as well. And so the Soviet accomplishment was personally confirmed by deputy NASA administrator Hugh Dryden.

The flight of the sixth Soviet lunar probe lasted thirty-eight hours, twenty-one minutes, and twenty-one seconds. On September 14, the Soviet scientists received the hoped-for confirmation from Professor Lovell that the space vehicle had stopped transmitting almost exactly at the predicted time. Luna 2 had stopped its transmissions one second later than calculated. And even this one-second difference was quickly explained, since it turned out that in calculating the probe's trajectory the ballisticians had not taken into consideration the elapsed time of the signal from the moon to Earth.

On September 14, 1969, the Soviet news agency TASS reported: "Today, on the 14th of September at 12:02:24 Moscow time, a second Soviet spacecraft reached the surface of the moon. Thus, for the first time in history, a space flight has been made from one celestial body to another. In memory of this remarkable event, a plaque bearing the emblem of the USSR and the inscription 'Union of the Socialist Soviet Republics, September 1959,' was placed on the surface of the moon. The reaching of the surface of the moon by this Soviet spacecraft is a remarkable success for science and technology. It is the beginning of a new phase in space exploration."

Mission Data, Luna 2	
Mission designation	Luna 2 (also Lunik 2)
Date	September 12, 1959, 08:40 CET
Spacecraft	E-1A no. 2
Booster rocket	8K72 (I1-7B) Luna
Spacecraft weight	860 pounds
Planned mission objective	Impact on the moon
Mission results	Impact on the moon
Last contact	September 14, 1959, 22:02 CET
Current location	Palus Putredinus, 0 degrees longitude, 29.1 degrees latitude

At 06:00 on September 15, this report was broadcast by every radio station in the Soviet Union, and a text with the same wording appeared in every morning newspaper. Korolev, Keldysh, and the other authors of the text were furious. It had gone through control and censorship committees and was supposed to be completely clean and free of any suspicious statements. But somehow a serious error had crept in, which caused the seemingly so successful Luna 1 mission of January to fall into disrepute. The passage in question read "a second Soviet spacecraft reached the surface of the moon." Of course, Luna 2 was in fact the first Soviet spacecraft on the moon. Because of a problem with its flight controls, Luna 1 had missed Earth's satellite. And now they had stated indirectly that the first probe was actually supposed to have come down on the moon.

Despite this, the mission was a stunning success, even though it was accomplished a year after the original target date. The Americans were still behind, and what was even better was that on September 15, Khruschev announced his long-planned visit to the United States, and therefore the success of Luna 2 was timed just right.

Now part 2 of the decree of May 28, 1958, had to be fulfilled. And this part was incomparably more difficult, since it required the use of a much more complex vehicle. For the first time in the history of spaceflight, a spacecraft had

to be created that could be controlled from the ground and also carry out its tasks independently with no input from Earth.

A camera unit was installed in the probe. As soon as the probe reached the designated area of the moon, the attitude control system had to turn the spacecraft so that the lens of the camera was pointed at the area of the moon that they wanted to photograph. This system's subsequent task was to keep the unit absolutely stable for forty to fifty minutes so that the photographs could be taken,

A suitable flight path had to be chosen. In particular, it had to take into account all the limiting factors it was subject to. The scientists decided on a highly elliptical orbit, at whose apex, 240,000 miles from Earth, the space probe would encounter the moon and fly around it. The probe did not have to carry out a steering maneuver; it was pure celestial mechanics. Insertion into this trajectory had to be very accurate, however. The probe would then take its photographs on the far side of the moon, at the apex of this steep elliptical trajectory. On the way back to Earth it would develop the pictures in developing fluid, dry them, and transmit them to Russian ground stations. The trajectory was chosen so that the distance from the moon would be about 4,200 miles.

Another problem was that the close flypast had to be made in such a way that the return flight to Earth came in over the Northern Hemisphere and had Crimea in its line of sight. The probe had to be as close to Earth as possible when it transmitted the photographic images; with the transmission methods in use at that time, picture quality was also a function of reception strength. At that time there were several Soviet antenna installations in Crimea that were suitable for "interplanetary" communication. They were on Koshka Mountain, near the famous spa at Simeiz on the extreme southern tip of Crimea.

The critical element on the probe was the Yenisey Photo Unit, which was manufactured by NII-380 in Leningrad. This facility would later make many more scientific optical devices for Soviet space travel. The camera was equipped with two lenses and could change exposure time automatically. The exposures began at a time determined by a command unit. It in turn received its information from the attitude control system. If the probe's position in space was stable and it was aligned correctly, the command was given to begin the photo session. The camera held a very limited quantity of chemical film material, and under no circumstances could it be wasted.

The system's mechanism was complex. As soon as the sequence had ended, during which it had to be wound frame by frame, the film was moved to a processing unit, where it was developed, fixed, dried, wound back into a special spool, and prepared for transmission.

A photoelectric cathode then turned the film negative into electrical signals. With the technical means then available, this was a very complex affair. It had to be scanned electronically, then a signal had to be created and enhanced and fed into the radio link. For the first time in the Soviet Union, semiconductors were used instead of transistors. At the time, all this was considered exotic and risky.

Even the radio link to Earth itself was a complex affair. It had to transmit not only the pictures to Earth but the telemetry as well, and the signals from the ground station to the spacecraft. All this—including the spacecraft itself—had been created in what by today's standards were incomprehensibly short time periods. To achieve this, however, testing had to be cut short in some areas.

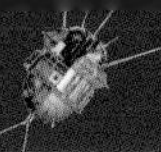

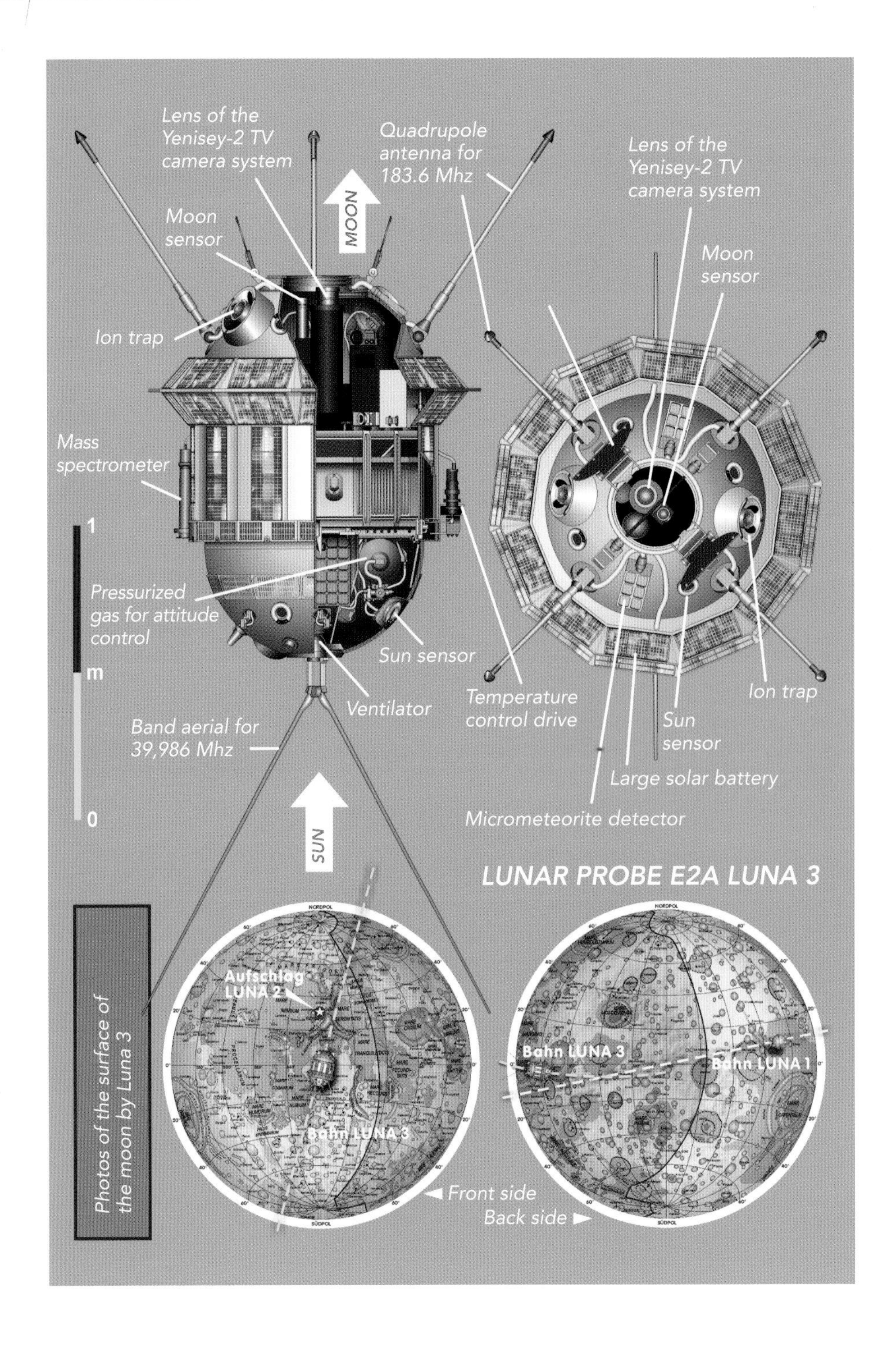

Lens of the Yenisey-2 TV camera system
Quadrupole antenna for 183.6 Mhz
MOON
Lens of the Yenisey-2 TV camera system
Moon sensor
Moon sensor
Ion trap
Mass spectrometer
1
Pressurized gas for attitude control
m
Sun sensor
Ventilator
Temperature control drive
Ion trap
Sun sensor
Band aerial for 39,986 Mhz
Large solar battery
0
SUN
Micrometeorite detector
LUNAR PROBE E2A LUNA 3
Photos of the surface of the moon by Luna 3
NORDPOL
Aufschlag LUNA 2
Bahn LUNA 3
Bahn LUNA 3
Bahn LUNA 1
Front side
Back side
SÜDPOL

Just twenty days after Luna 1, the probe Luna 2 was launched on October 4, 1959—the second anniversary of the beginning of the space age—to photograph the moon. As had become normal after all the failed launches, TASS cautiously announced that the launch had been a success but not the mission objective.

This decision proved a wise one, because communications with the spacecraft soon worsened. Telemetry from Luna 2 barely reached the station in Crimea, and the probe itself was obviously not receiving any signals from Earth. The antenna personnel in Koshka called for help, and so the experts of the OKB-1, especially Korolev, Chertok, and Arkady Ostashev, the specialist in the probe's radio equipment, flew to Crimea as quickly as possible to see what was happening. For this purpose, a Tupolev Tu-104, the USSR's first jet airliner, which at that time was available only in small numbers, was put at their disposal.

The aircraft landed at a military airfield near Koktebel. There they boarded a helicopter whose rotor blades were already turning. It was supposed to take them to Ay-Petri in southern Crimea. The plan called for an automobile to pick them up there and drive them to the station at Koshka. Soon after the take-off from Koktebel, the helicopter's commander entered the cabin and informed the passengers that it was snowing heavily in Ay-Petri and that a landing there was not advisable. The next best safe landing place was Yalta. So the flight went to Yalta, where it was met by the local party secretary. He placed an automobile at their disposal, and after several detours they finally reached the village of Simiez at the foot of Koshka Mountain, where the control center was located. Also there was the equipment with which the hoped-for pictures from Luna 2 were to be printed on heat-sensitive paper, which needed no photographic development. The images were also transferred to movie film at the same time. It could not be developed on the spot, however, and first had to be taken to Moscow.

Korolev and his people were briefed first of all. It turned out that the shape of the antenna was responsible for the poor communications. There was nothing they could do about it. But at a glance, Korolev also realized that something wasn't right about the command structure in the control center. There were two senior officers in the room, along with the facility's technical director, Yevgeny Boguslavsky, a civilian. With his unfailing intuition when it came to people, Korolev also discovered that all three operators were giving commands, which were not coordinated. All were turning the hundreds of knobs and throwing switches, but the situation was not improving. In a matter of minutes Korolev set up a clear command structure, gave the two officers other tasks, and directed the personnel to henceforth listen only to Boguslavsky's commands.

And it worked. Data transmission from the probe, which was on its way to the moon, began at about 18:00 on October 6. The telemetry data came in, and it was clear that everything was in order on the probe. They were to carry out their photo session on the morning of October 7, then develop the pictures onboard and finally transmit them to Earth from a distance of about 30,000 miles.

The battery-powered probe's transmitter was a very low-power system, and any external interference had to be eliminated. The Soviet Black Sea Fleet was therefore directed to halt all radio traffic at that time, so as not to interfere with the weak reception signal. At the last minute it turned out that too little of

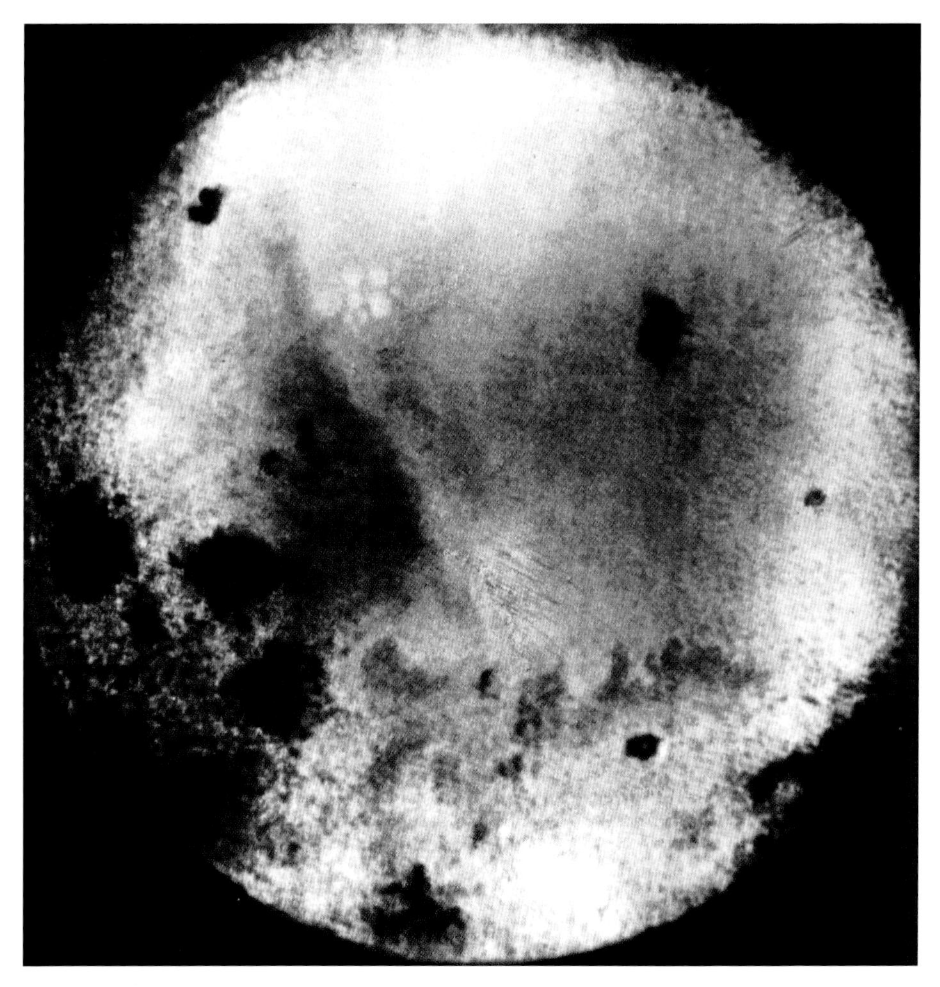

The quality of the images of the back side of the moon was less than perfect; nevertheless, the most-prominent features could be easily recognized, such as the Mare Moscoviense (top right) *and the Tsiolkovsky crater* (bottom right).

the special magnetic tape, with which they were supposed to secure the incoming data, was available on-site. A jet aircraft had to be organized to bring in more at the last minute. The magnetic tape arrived at the last moment before the data were received.

Unlike today, only one throughput was available, and there was therefore exactly one chance. Performance of the onboard batteries was limited, and the scientists were not yet sure that the probe would not burn up in Earth's atmosphere on the descending branch of its long orbital ellipsis.

The next day, at 06:30 Moscow time, at the apogee of its orbital ellipsis about 4,200 miles from the moon, Luna 2 began taking photographs. Prior to this the engineers in the control center had turned off the telemetry

transmission system to save battery power. After about forty minutes the photographic sequence was completed and the complicated process of developing the film inside the probe began.

Many hours later, the call finally came from the control center: "Range 30,000 miles. Signal stable. We have reception." And on the heat-sensitive paper of the receiver system, the first photo of the far side of the moon appeared, line by line. The quality was very poor. One could see the curvature of Earth's satellite and, with some imagination, several dark spots. That was rather disappointing. The following pictures were better, but not very much so. Not until the exposed film was developed in Moscow several days later did more details appear, enough to recognize about a dozen topographical features. The task of naming these terrain features now fell to the Soviet Academy of Sciences. It is to the credit of the academy members surrounding Mstislav Keldysh that the people honored by being named included many non-Soviet scientists, including Giordano Bruno, Jules Verne, Heinrich Hertz, James Maxwell, Marie Curie, and Thomas Edison. The list did, of course, include persons such as Nikolai Lobachevsky, Dimitri Mendelayev, and Alexander Popov.

November 26: Pioneer 5 (also Pioneer P-3) was planned as an orbital lunar probe, but the launch atop an Atlas-Able booster rocket failed because the payload fairing collapsed forty-five seconds after liftoff.

Mission Data Luna 3	
Mission designation	Luna 3 (also Lunik 3)
Date	October 4, 1959, 01:43 CET
Spacecraft	E-2A no. 1
Booster rocket	8K72 (I1-8) Luna
Spacecraft weight	615 pounds
Planned mission objective	Photograph the dark side of the moon
Mission results	Photographed the dark side of the moon from a distance between 39,457 and 41,445 miles and, on October 6, transmitted twenty-nine photographs, including seventeen of the Earth
Last contact	October 22, 1959
Current location	Entered Earth's atmosphere on April 29, 1960

1960: FIRST HALF OF THE YEAR

The space race between the Soviet Union and the US to be the first nation to put a man on the moon had not yet fully begun. The focus at the beginning of 1960 was the achievement of other firsts, and so after this first series of lunar probes, attention centered on manned orbital flights and the first probes to Venus and Mars. The point of view in 1960 was that there was still time for the moon. At that time the Americans were not yet planning any soft landings by unmanned lunar probes. The Apollo project was at a very early stage, and it did not look as if America would strive for a moon landing before the mid-1970s. Kennedy's great programmatic speech, with which he committed the Americans to a manned moon landing before 1970, was still a year and a half in the future. In the Soviet Union, however, the OKB-1 was already planning the next generation of unmanned lunar vehicles, the E-4 landing probes. They were confident that they could again beat the Americans; however, they did not want to nor could they assign special priority to this project. Preparations for the manned orbital missions were more urgent, since the USSR suspected that the US wanted to put a man into orbit in 1961.

Also important were the first probes to Venus and Mars, because there too the Americans were still in the starting blocks. Both planets were still completely unknown. In 1960, it was still thought that Venus might be a cloud-shrouded, life-sustaining tropical world, and since Orson Welles's radio broadcast of H. G. Wells's *War of the Worlds*, Mars had been regarded as a potentially life-supporting planet. They could come back to the moon later.

Mstislav Keldysh—like many others—was, however, dissatisfied with the quality of the photographs from Luna 3. It was clearly apparent that the image sharpness delivered by the camera system was nowhere near what could be achieved on Earth. In addition to the radio equipment on the probe, this was mainly due to the antenna system of the receiving station in Crimea. If the background signal noise could be filtered out more effectively, the result would be higher-resolution photos.

The improved antennas were almost complete at the beginning of 1960. They had a larger receiving surface, more-powerful noise suppressors, and receivers with ten times the capacity of the previous equipment.

Despite Korolev's insistence on putting off a repetition of the circumlunar flight, pointing out that the 1960 flight program was already overcrowded, Keldysh was not ready to compromise. He managed to obtain a government decree ordering another circumlunar mission to be carried out in April 1960, in order to obtain better images of the far side of the moon. In the end, two 8K72 rockets were readied for the mission, and in March two hastily built flypast probes were also delivered to Baikonur. They were given the designation E-3. After the atomic impact probe had finally been put to bed, this number had become available.

On April 12, after much overtime and many night shifts, the rocket with the E-3 probe was ready for launch in the assembly hall at Baikonur, and ready for transport to launchpad 1/5. Despite significant problems during the checkout of the rocket and payload, during which time several systems had to be completely or partially replaced, the launch took place on April 15, 1960, exactly at the time selected months earlier: 18:06:42 Moscow time. At first all proceeded normally—ignition of the engines, the first launch phase, separation of the booster, jettisoning of the payload fairing, ignition of the third stage, and jettisoning of Block A. But then, suddenly, premature burnout. The engine had shut down three seconds too soon. The rocket's speed was 425 feet per second below what was required to reach the moon.

The investigation, which was launched immediately, revealed something dreadful. There had been no technical failure. The people under Vladimir Barmin who were responsible for carrying out the launch had literally overlooked

Mission Data, E-3 No. 1	
Mission designation	—
Date	April 15, 1960, 16:07 CET
Spacecraft	E-3 no. 1
Booster rocket	8K72 (I1-9) Luna
Spacecraft weight	617 pounds
Planned mission objective	Photography of the dark side of the moon from a closer distance than Luna 3
Mission results	Achieved velocity too low for translunar injection, since the upper stage was not fully fueled. Achieved distance from Earth of 124,274 miles

fully fueling the Block E stage of the rocket. It was an embarrassing incident, but there was no time to think about that, since the second rocket with the replacement probe was in the assembly hall, ready for launch.

The events that followed are worth citing Boris Chertok's account from his biography word for word—no one else was as close as he to the R-7 8K72 with the serial number L1-9A, which was supposed to launch the second E-3 probe on a circumlunar flight in the early evening hours of April 19, 1960. He wrote:

> After three restless days, on April 19th, the next rocket with the second E-3 lunar probe was ready for launch. This time I wanted to take advantage of the evening twilight, and at T-15 minutes I decided to leave tracking station IP-1 [author's note: tracking station IP-1 was about 0.6 of a mile from the launchpad], which was filled with numerous onlookers, and go outside to experience the launch outdoors. I took my time, taking in the smell of the steppe, walked about 1,000 feet in the direction of the launchpad, and gazed at the rocket, which was bathed in bright light.
>
> I was able to hear the announcement "T-1 minute" over the loudspeaker. There, out in the steppe, I was surrounded by a feeling of loneliness; there was no one nearby. There was only the picture of this wonderful dream, which was embodied in this rocket. At that moment I thought: "If something happens to you now, then there is nothing that I and hundreds of your other creators can do for you." And then it happened. My thoughts had surely summoned the disaster. All of the rocket's engines in the main stage were running and produced an ear-shattering noise. As I was standing 1,000 feet closer to the rocket than anyone else, I clearly felt the difference in the noise level [between the preliminary stage and the main stage].
>
> What's happening now? I surmised, more than I saw, that the booster facing me was not lifting off with the other boosters and the main block, and that it was collapsing into itself, spewing flame. The remaining three boosters and the main block hesitantly lifted off, and just as they were flying directly over my head the whole combination fell apart. I could not say what was flying where, but I felt that one of the boosters was heading straight toward me, engines roaring. Run! RUN! To the IP, where there were safety trenches. Perhaps I might make it. I had been quite a good 100-meter runner in my Comsomol days, the best sprinter in Factory 22. Now, here in the steppe, with this booster heading toward me, I probably set my personal record.
>
> But the steppe is no running track. I stumbled and fell and came down hard on my knee. Behind me there was a tremendous explosion, and I felt the shock wave of hot air. Chunks of Earth thrown up by the explosion fell to the

Mission Data, E-3 No. 2	
Mission designation	—
Date	April 19, 1960, 17:08 CET
Spacecraft	E-3 no. 2
Booster rocket	8K72 (I1-92) Luna
Spacecraft weight	617 pounds
Planned mission objective	Photography of the dark side of the moon from a closer distance than Luna 3
Mission results	The booster rocket's Block B booster achieved only 75% of nominal performance during launch. The rocket reached a height of only 600 feet, then broke apart and severely damaged the launchpad

> ground all around me. I clenched my teeth and hobbled back to the tracking station, away from that enormous wall of flame that was now spreading at the spot where I had just been standing. But where were the other boosters? A pillar of fire rose from the MIK [vehicle assembly building]. I feared that one of the booster blocks had struck the assembly hall. There were many people inside.

The accident caused enormous damage, but fortunately there was no loss of human life. It had been the central stage that came down right next to the MIK. All the windows and doors were blown out of the building, and the concrete floor was reduced to fragments and crumbs, but only one person in the building, an officer, was injured, sustaining bruises.

1960: SECOND HALF OF THE YEAR

After these two failed attempts to undertake another lunar photo mission, flights to Earth's satellite were interrupted until January 1963. During this time, development of the E-6 series of landing and orbital probes was on the back burner. As far as the civilian space program was concerned, OKB-1 was concentrating on the manned missions and the first attempts to reach Venus and Mars.

The manned flights in particular were a matter of national pride. After the success of Sputnik 1, leaving it to the Americans to put the first man into space was completely unthinkable. Ultimately, proof of the superiority of socialism was at stake. Generally speaking, the manned flights of the Vostok and Voshkod programs cannot actually be attributed to the USSR's lunar program. They were simply firsts. There was almost no scientific equipment in the cabins of the Vostok spacecraft. Its purpose was to complete at least one orbit of the Earth and survive. Not until 1967, with the orbital version of the Soyuz, was there a spacecraft that in principle could be used universally (and it is to the present day) but was conceived primarily as the spacecraft for the manned Soviet flights to the moon.

A four-stage version of the R-7 was planned for the planetary missions, but also for the flights to the moon by the E-6 probes, which were supposed to carry out the first soft landings and orbital missions. This version of the rocket was given the article number 8K78. This fourth stage was designated Block L, in keeping with the next letter in the Cyrillic alphabet after *Ye* (or E). The rocket was later given the name Molniya because it placed a large number of communications satellites with the same name into orbit.

The interplanetary spacecraft of the Venera and Mars series would each weigh 0.55 tons. The E-6 probes would weigh 1.5 tons. The payload capacity of the three-stage 8K72 Luna was much too small for these launch feats. As well, the inability for reignition in orbit was a handicap that prevented a precise trajectory injection. Calculations by the ballisticians had revealed that the most effective injection method for a trajectory to the moon or the planets required the first three stages of the R-7 to initially achieve a parking orbit. This would be followed by a free-flight phase for the orbital unit, consisting of the fourth stage and the space probe, which guided it to the injection window, a previously

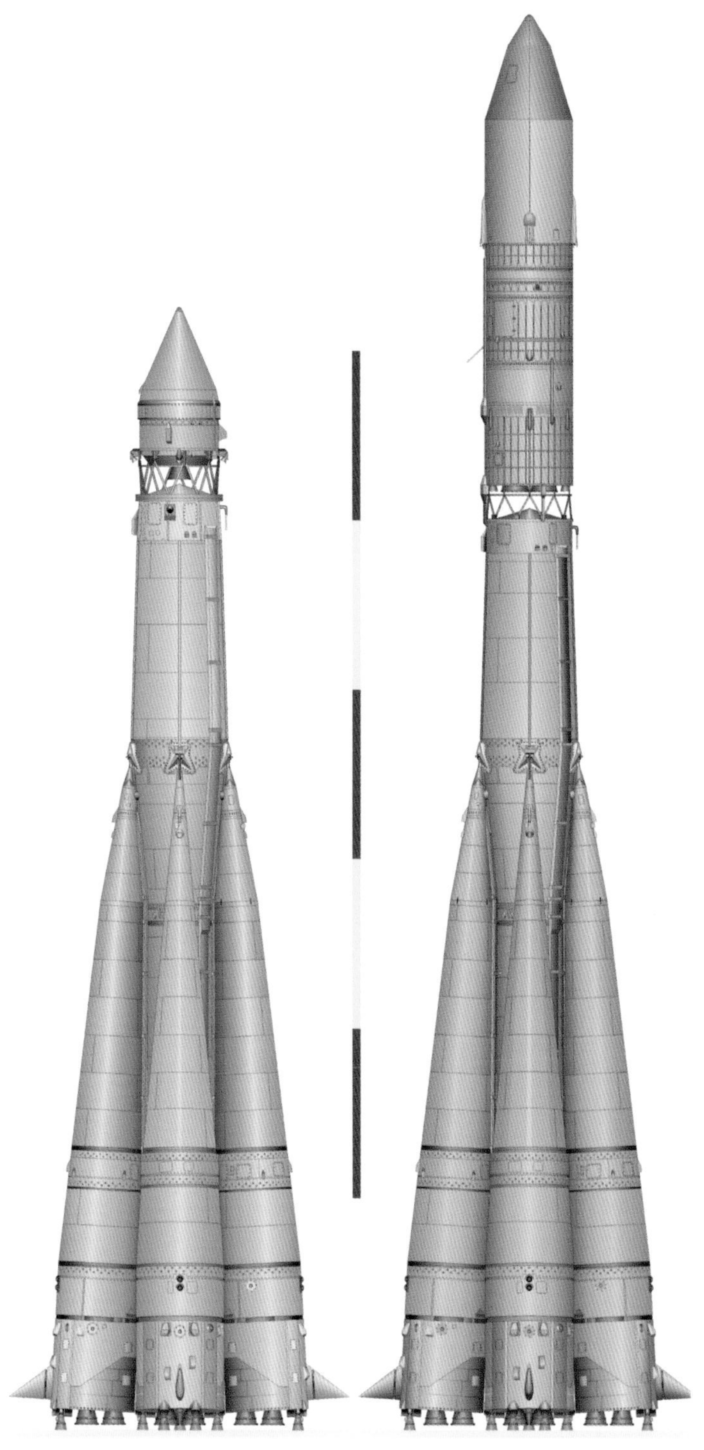

Comparison of the 8K72 Luna with the 8K78 Molniya.

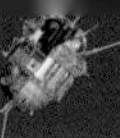

calculated optimal point in the orbital trajectory at which the fourth stage had to be ignited to achieve the translunar or heliocentric trajectory.

In the time between burnout of the third stage and ignition of the fourth stage, the combination was in free fall, or in the condition of weightlessness. This caused the technicians some headaches, since Block L had to be ignited during this weightless phase. The fuel in the tanks was of course also weightless, and a procedure had to be devised that caused the fuels to return to the bottom of the tank and ensure the flow of fuel and oxidizer to the combustion chamber. This was achieved with a small solid-fuel engine, which was put into operation just prior to the ignition of Block L. The fuel in the tanks experienced acceleration, and ignition could be carried out. The disadvantage: afterward, the solid-fuel rocket was spent and could not be ignited again.

As had been feared, it took a long time for the new rocket's technical problems to be ironed out. Its first uses, on October 10 and 14, 1961, were supposed to have seen the first two Mars probes sent on their way to the red planet. Both times, however, the RD-108 four-chamber engine in the third stage failed. The new Block L fourth stage was not used. The third flight, with the first Soviet Venus probe, also failed on February 4, 1961. This time the Block L stage failed to ignite and the combination of stage and spacecraft was stranded in orbit. Not until launch no. 4 on February 12, 1961, with a second Venus probe onboard (which after the successful launch was named Venera 1), was the 8K78 successful for the first time.

September 25: Pioneer P-30 was planned to be a lunar orbiter and was supposed to enter lunar orbit sixty-two hours after launch. However, the second stage of the Atlas-Able booster rocket failed during launch, and the third stage and its payload crashed into the Indian Ocean.

December 15: On this day, another American mission with the objective of lunar orbit failed. An Atlas-Able 5B rocket was supposed to place the space probe Pioneer P-31 onto the transfer trajectory, but it exploded sixty-eight seconds after leaving the launchpad at an altitude of 7 miles above the launch site. The cause was found to have been structural failure in the adapter between the first and second stages.

1961

In the aftermath of the Vostok 1 mission, President John F. Kennedy asked Vice President Lyndon Johnson what possibilities there were of restoring America's technical prestige with an outstanding space first. This question started the American machinery running. The Apollo program was waved through in a crash effort involving the top NASA managers, White House staff, and members of Congress. On May 25, 1961, President Kennedy gave his now-famous speech to both houses of the US Congress, in which he spoke about the future of American spaceflight and the US position in this discipline compared to its great rival, the Soviet Union. The last four sentences of this speech have become world famous and

continue to be quoted up to the present day. Kennedy said, "we cannot guarantee that we shall one day be first, but we can guarantee that any failure to make this effort will make us last. We take an additional risk by making it in full view of the world, but as shown by the feat of Astronaut Shepard, this very risk enhances our stature when we are successful. I believe this nation should commit itself to achieving the goal, before this decade is out, of landing a man on the moon and returning him safely to Earth. No single space project in this period will be more impressive to mankind, or more important for the long-range exploration of space, and none will be so difficult or expensive to accomplish."

When Kennedy made his speech, the total US experience in manned spaceflight was the fifteen-minute-long ballistic "space hop" by Alan Shepard in the Mercury capsule Freedom 7. While the flight was spectacular and dangerous, it was anything but a groundbreaking demonstration of the superior spaceflight capabilities of the United States. At the time of Kennedy's speech, the first American manned orbital mission was almost a year in the future. And yet the president had called for nothing less than landing men on the moon and bringing them safely back to Earth within the next eight years—a space journey by three astronauts lasting two weeks to a celestial body 240,000 miles away from their home planet. With this announcement, Kennedy had thrown down the gauntlet at the feet of the Soviet Union. The race to the moon began in all seriousness that day. The US Senate and House of Representatives agreed without opposition to the massive increase in NASA's budget. At first this speech scarcely registered in the Soviet Union and was dismissed as political rhetoric with no real background.

Three months later, the American position looked even worse when on August 6, German Titov departed on a flight lasting more than twenty-four hours, with Vostok 2 circling the Earth eighteen times. That was roughly equal to the number of hours the Americans planned to spend in space by the end of 1963. Titov survived the flight unharmed, although the Vostok's old problem reappeared, and during its return to Earth the equipment section failed to release immediately from the spherical spacecraft. Like Gagarin before him and all future Vostok cosmonauts, Titov landed by parachute, having previously ejected from his capsule, which was under its own parachute. The capsule landing speed of more than 30 feet per second required this method if they wished to avoid injury to the cosmonauts during landing. This very nearly happened, however, since Titov came down just 30 feet from a railroad track on which a train passed by at that very moment. The landing took place near the small town of Krasny Kut in the Zaratov District, not far from the Kazakh border, which at that time was of course not a border, since Kazakhstan was part of the Soviet Union.

Chief designer Oleg Ivanovsky of OKB-1 escorts Yuri Gagarin to his spacecraft prior to his historic flight.

The committee that evaluated the spaceflight recommended that a representative of the state railway be included in future launch decision-making processes, so that the train schedules could be coordinated with the landing times of the cosmonauts.

Apart from this and other minor mishaps, the flight was a complete success. Titov was not only the first human who spent more than a day in space, he was also the first man to sleep in space. He was also the first to experience space sickness, which the Soviets tried to keep secret at the time.

August 23: The space probe Ranger 1 was supposed to be placed in an extremely high Earth orbit in order to test the systems and flight strategy for future Ranger lunar probes. The strived-for orbit had a perigee of 36,000 miles and an apogee of 680,000 miles. Instead of this, Ranger 1 remained in a low parking orbit around Earth, because the second ignition of the Atlas Agena B booster rocket failed. Ranger 1 burned up in Earth's atmosphere on August 30.

1962

January 26: Ranger 3 was dispatched on a lunar-impact mission. The launch was faultless, but navigation and control problems developed on the way to the moon, and as a result the probe missed the moon by 22,800 miles.

February 20: John Glenn made the first orbital flight by an American, circling the Earth three times in Friendship 7.

April 23: Ranger 4 was dispatched on a lunar-impact mission. Once again the launch was faultless, but because of a malfunction of the onboard computer the solar generators failed to deploy and the navigation system was not activated. The batteries ran out ten hours after launch. On April 26, Ranger 4 crashed on the reverse side of the moon after a flight time of sixty-four hours.

May 24: Scott Carpenter repeated Glenn's flight and circled the Earth three times in Aurora 7.

At the end of February 1962, training began for a joint flight by two manned spacecraft. At Korolev's suggestion, the mission was intended to be a powerful response to Glenn's flight, demonstrating to the Americans how hopelessly behind the Soviet Union they were. The air force selected Andrian Nikolayev and Pavel Popovich as the two cosmonauts who were to carry out the mission. Valery Bykovsky and Vladimir Komarov were selected as their backups. The launches were to be just one day apart. This was something

Pavel Popovich in Vostok 4.

new and required that the launch crews work almost around the clock for three days. The first day would be preparations for Nikolayev's launch in Vostok 3; the second day, the launch and the preparation of Popovich's rocket; and on the third day, the launch of Vostok 4.

As soon as the two spacecraft approached one another, the two cosmonauts were supposed to establish radio contact. Korolev and Keldysh wanted a flight duration of three days for each spacecraft. The man responsible for training the cosmonauts, General Nikolai Kamanin, insisted vehemently that each man spend just twenty-four hours in space, but he agreed to extend each flight by one day provided the cosmonauts were found to be in good condition. Finally they agreed to return Nikolayev after four days during the sixty-fifth orbit, and Popovich after three days and forty-nine orbits. The final decision would not be made until the mission was already in progress.

And so, Vostok 3, carrying Nikolayev, was launched on August 11, and Vostok 4 took off with Popovich on August 12. The uncertainty about the flight duration became a little grotesque. On the first day of the flight, the question of mission duration had still not been settled, and there was no one who wanted to make the decision. Korolev refused, as did Marshal Rudenko and even Marshal Grechko. The question was finally put to Nikita Khruschev himself, who Solomon-like passed the following judgment: "If there are no problems with the spacecraft or Popovich's health, then simply ask him if he wants to. And if he can fly longer, then he shall complete the four days."

Korolev, Kamanin, and Gagarin therefore spoke with Popovich and asked him how he felt. He replied, "Quite excellent." The result of this conversation was passed on to Khruschev by the secretary of the party central committee, Frol Kozlov. Thus the decision was made for four days. Or at least it should have been made, because at the beginning of the third orbit there was a technical problem. The temperature and humidity onboard fell to the lowest acceptable value. The temperature inside the spacecraft was just 50 degrees Fahrenheit. At that point in the mission, the doctors became uneasy and demanded that Popovich land immediately.

What now? They had just asked the highest-ranking party leaders to authorize the four days, and suddenly they were supposed to tell them that three days were not even possible. The situation would become even worse when Popovich sent a curious message to Earth. Accompanied by heavy static, his message came from orbit: "I am observing a thunderstorm." "Thunderstorm" was the agreed-upon code word for space sickness. The Russians used such code words to avoid betraying possible problems to Western observers listening in on their radio traffic.

The uneasiness in the control center became even greater, until Popovich himself finally cleared up the matter. During the next radio contact he was asked about the "thunderstorm" and amended his report to "I am observing a meteorological thunderstorm and lightning."

In the end the flight was concluded as planned. Nikolayev landed after four days and six minutes, while Popovich landed after two days, twenty-two hours, and fifty-seven minutes. Once again the Soviets had left the Americans far behind.

October 3: Walter Schirra launches in Sigma 7 and completes seven orbits of Earth.

October 18: The Ranger 5 mission is also a failure. A short-circuit in the energy supply system cuts the power lines from the solar generators to their users onboard the probe. Because of the lack of power, a hastily initiated course correction cannot be completely carried out. Eight hours and forty-four minutes after launch the spacecraft was dead. It missed the moon by 450 miles.

1963

Until the beginning of 1963, the Soviet leadership had no cause to worry about the Americans. They were obviously years ahead everywhere in space. One Soviet first after another shocked the frustrated citizens of the US, whose understanding was that the Soviets had rushed out to an almost insurmountable lead. American accomplishments in space to date gave no reason to think that they would be in the position to in fact put men on the moon by the end of the decade, as they had promised to do.

Beginning in the spring of 1963, however, there began a slow change of heart. In the Soviet Union there were increasing signs that the Americans were serious about their lunar plans. By that time, NASA had completed four perfect missions with the Saturn Block 1. The basic design of the Apollo had been frozen in 1962, and in March 1963 the contract for the lunar lander had been given to Grumman. As well, NASA had decided on the lunar orbit

rendezvous as the preferred mission mode for Apollo, and for the transport of all elements to the moon on a single Saturn V rocket. All these ever-clearer signs began to gradually unsettle the Soviet leadership.

That summer, at the behest of Nikita Khruschev, Korolev was working on a moon flight strategy that they could beat the Americans with. The key role in this scenario would be played by the new N1 large booster rocket, which had to be developed. Korolev's plan consisted of five program sectors with manned and unmanned elements. They all began with the letter *L* for luna, or moon. Their distribution was as follows:

L1 *Circumlunar missions on highly elliptical Earth orbits*

L2 *Unmanned lunar orbiter and lander to prepare for and support the manned landings*

L3 *Manned moon landings*

L4 *Second-generation manned lunar orbiter*

L5 *Advanced manned lunar rover*

The abbreviated designations for this five-phase program soon established themselves in Soviet spaceflight terminology. In the years that followed, they were used for the affected program elements independently of the products of individual design bureaus. In practice only the L1, L2, and L3 phases took effect, while L4 and L5 remained studies.

The manned circumlunar flights were supposed to begin in the mid-1960s as part of the L1 program sector. Development of the N1 would obviously not be completed by then, and therefore it was necessary to transport the components of the circumlunar spacecraft and its fuel into orbit by using smaller rockets. There the spacecraft would be assembled and fueled before it could make its way from low Earth orbit to its highly elliptical trajectory. The spacecraft was to be based on a variant of the Soyuz spacecraft then under development by OKB-1 and the booster rocket of the same name, a developed version of the R-7. A total of six Soyuz rockets would have to deliver payloads to orbit for a single circumlunar mission.

The L2 subject area was to comprise orbital flights and landings on the moon by unmanned spacecraft. The landing craft were initially supposed to deliver stationary landers, and in a later phase also rovers, which would scout landing sites for the L3 phase. The first phase of this program was to be accomplished with 8K78 booster rockets.

The L3 phase of the program was to see the first missions by the N1, initially unmanned and then manned. With its payload of 77 tons, according to this plan a single N1 was supposed to be capable of carrying out a lunar orbital flight with a modified Soyuz. As soon as they had achieved safety and gained experience near the moon, and the unmanned preliminary missions had been carried out safely, the next phase of the program would consist of unmanned and then manned landings. They were supposed to begin in late 1967 or early 1968.

In Korolev's scenario of early autumn 1963, they would require three N1s and one Soyuz per mission. The first N1 was supposed to place a complete but unfueled moon flight complex into Earth orbit. Two more N1s would follow with fuel, which then had to be transferred to the moon ship automatically. Finally, a Soyuz booster rocket carrying a Soyuz spacecraft would be launched. It would dock with the lunar flight complex, and afterward the actual moon-landing expedition could have begun with insertion into the transfer trajectory to Earth's satellite. In orbit, the entire complex would have weighed 220 tons. In the end, 23 tons of this would have landed on the moon. Exactly like the American Apollo program, the crew would have consisted of three cosmonauts. Two of them would have landed on the moon while one remained in orbit in the mother ship.

This original mission plan was modified several times in the coming years. The most radical change took place in August 1964. From then on, the entire moon-landing combination was to be transported by a single rocket, similar to the American plan. The reasons for this abrupt change were manifold, but in the end it was economic and logistical considerations that were decisive. The idea definitely did not come from Korolev and Mishkin. Rather, it was the Soviet leadership, which looked to the American example and directed the engineers to proceed in exactly the same way. And of course this scenario was also much cheaper, since it required a single rocket instead of the previous four. However, this decision opened a Pandora's box of problems, which were unleashed by a single fact. The N1 rocket was far from powerful enough for the single-launch scenario. One could pare the landing procedure as much as one wanted, but the payload was and remained much too small. The N1 could transport only 82.7 tons into low Earth orbit. By comparison, this figure for the Saturn V was almost 155 tons, and even the Americans were struggling with weight problems. A study by OKB-1 revealed that a lunar landing using the single flight mode would require a weight of 105–110 tons from low Earth orbit. OKB-1 had originally planned a crew of three like the Americans: two men for the moon landing and one circling the moon in the mother ship. It was now clear that this would no longer work for a single launch. The problem had to be approached from two sides. The mission equipment had to be pared down and the rocket had to become larger.

As a result, the L3 lunar combination was defined in 1964. It consisted of the following four elements:

- *Block G. Essentially the fourth stage of the N1. It was envisaged solely for the translunar acceleration maneuver.*
- *Block D. Its purpose was course corrections en route to the moon and for braking into lunar orbit. In addition, it would be responsible for the major part of the descent to the surface of the moon.*
- *The LK (Lunniy Korabl) lander, and the*
- *LOK (Lunniy Orbitalny Korabl) lunar orbital vehicle, or the "mothership."*

The crew was now reduced from three to two. A single cosmonaut would land on the moon while his colleague circled the moon in the mother ship. A heavy docking system with entry tunnel between the lander and mother ship was also deleted. The lone cosmonaut would now have to undertake a "space walk" to reach the lander. With this and many other measures, the designers attempted to reduce the weight of the lunar combination sufficiently that it matched the payload limits of the N1 rocket.

But back to the year 1963, when the first experiments for an unmanned lunar landing began. This program was initially proposed by Sergey Korolev, and on December 10, 1959, it was ordered by a decree issued by head of state and party Khruschev. Here too the primary objective, a soft landing on the moon, was in any case to be achieved before

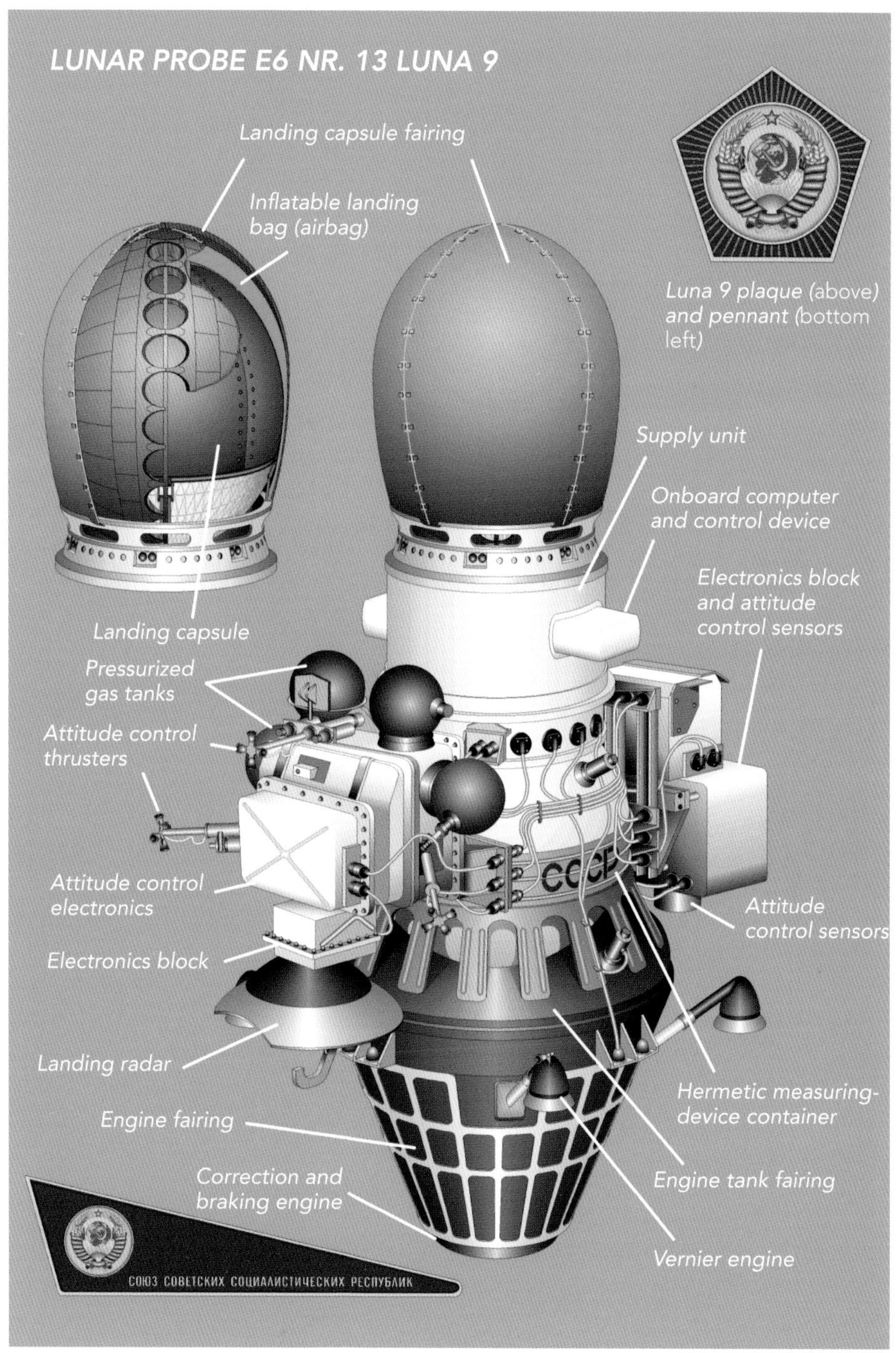

Lunar probe E-6 no. 13 Luna 9.

the Americans, and photos and physical data were to be transmitted to Earth from the surface of its satellite.

OKB-1 initially gave the plan little priority. The other programs seemed too important, especially the manned flights, the missions to Venus and Mars, the Zenit reconnaissance satellite program, and the R-9 ICBM, to allow significant resources to be allocated to the moon-landing plan. The Soviets watched the Americans and their Surveyor program closely, and it was obvious that several years would pass before the Americans could handle such a project.

But in reality the Soviets did not have as much time as they supposed. The problems involved in achieving the astronautical tasks had been severely underestimated. Not all the design parameters for the probes had been determined by the end of 1961, to say nothing of initiating completion. It soon turned out that all of the spacecraft's control mechanisms represented a particularly difficult problem. The spacecraft had to be precisely aligned for the landing, altitude and speed had to be precisely maintained so that the landing engine could begin its braking burn at the correct second, and speed had to be completely eliminated by the time it reached the surface of the moon. For this a radio altimeter had to be developed, basically a simple radar that passed its information to the engine control unit. This was no simple design problem in the age of analog flight control with no onboard computer. Many orders had to be transmitted from ground stations to the probe at exactly the right time to initiate mainly mechanical processes controlled by perforated tapes. There were no variable-thrust engines at that time, and the precise speed changes for the last several feet prior to landing had to be accomplished with small thrusters.

It was clear that the previously used 8K72 variant of the R-7 was unsuitable for the lunar soft-landing project. It could place only 715 pounds on a lunar transfer trajectory. This brought into play the 8K72, with which the Soviets were just beginning to carry out flights to Venus and Mars. In addition to the fact that its performance was also at the absolute lowest limit, it also had another decisive disadvantage: it was extremely unreliable. In the two years prior to the E-6 program, the 8K78 was used ten times for the Mars and Venera program. Only two of these launches were successful: the missions of Venera 1 on February 12, 1961, and that of Mars 1 on November 1, 1962. Of the eight failures, five were the fault of the fourth stage.

From the beginning, the weight of the probe was extremely problematic. Even with the simplest design the landing capsule alone weighed 220 pounds. The engineers determined that the entire probe, even with the most-drastic weight-saving measures, could not be brought below 3,306 pounds. Even the control unit with the designation I-100 weighed more than 175 pounds. But to maintain this weight, decisions had to be made about this vital piece of equipment, which would very negatively affect the program. For weight reasons, the control elements for the third and fourth stages of the booster were combined in the I-100, and redundancies were also largely done away with. Even with these risky decisions, however, the weight of the entire combination ultimately rose to 3,483 pounds.

Mstislav Keldysh assigned his best people from the Academy for Applied Mathematics to calculate trajectories with which the limited performance of the rocket could best be used. Finally, the mission profile of the E-6 probes was as follows.

The first three stages of the R-7 8K78 initially placed the combination of the E-6 probe

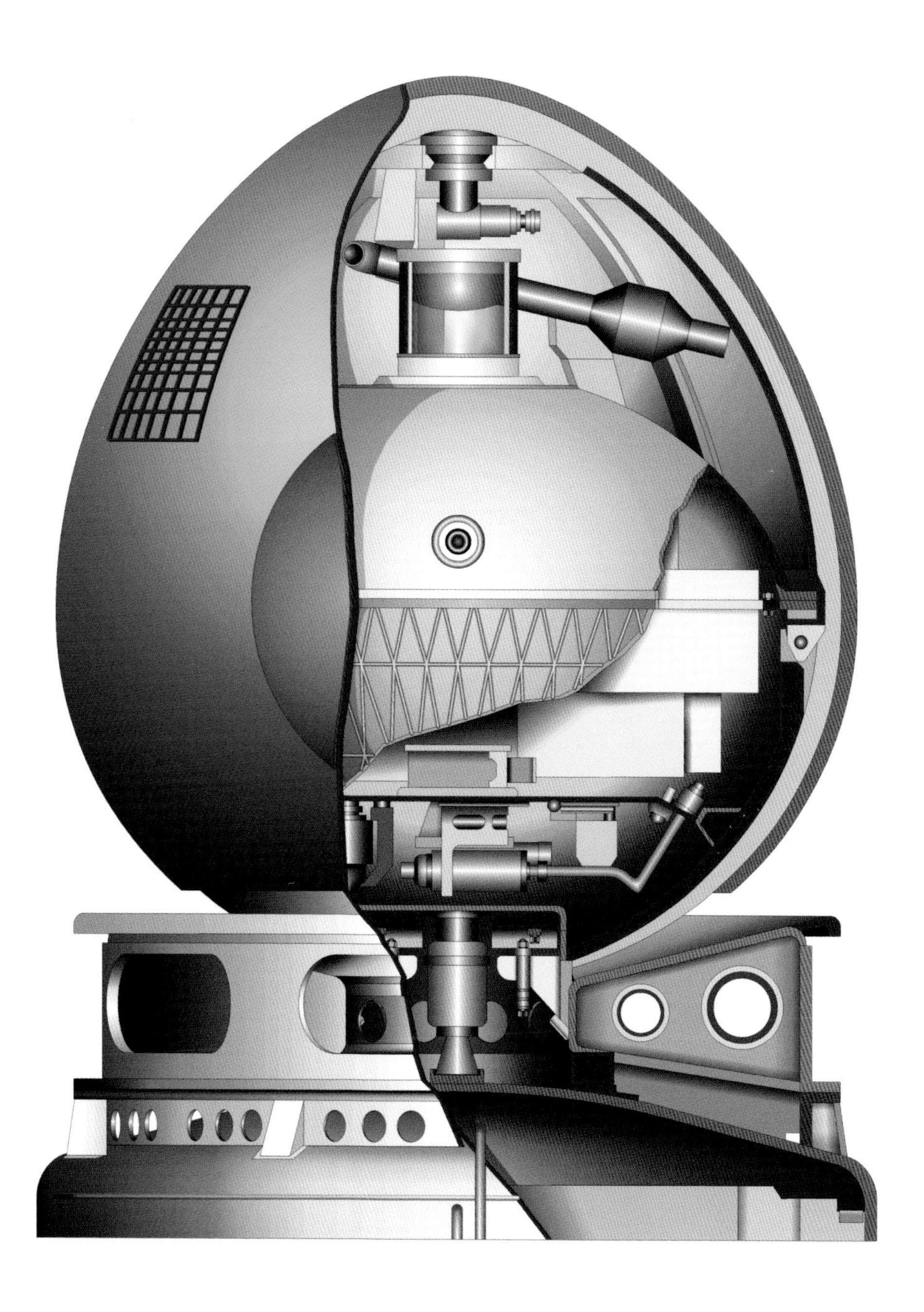

Landing capsule of the E-6 probes Luna 4 to Luna 14.

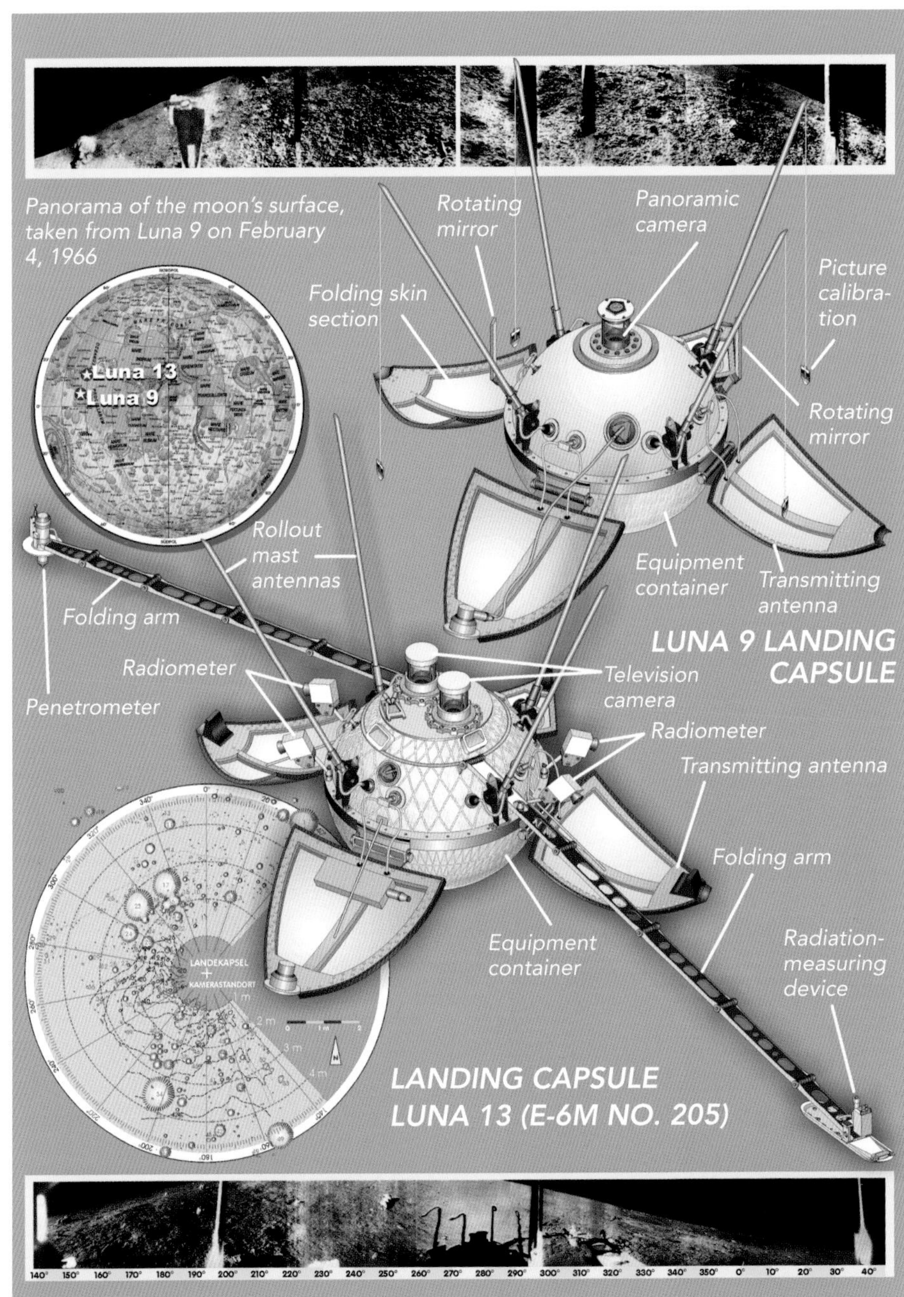

Luna 13 panorama no. 2 of the moon's surface, taken on December 26, 1966, between 16:00 and 18:23 Moscow time (above, to the left, is a map of the landing site).

and the Block L stage into a low Earth orbit. Its perigee was about 105 miles, its apogee was about 138 miles, and inclination of the orbit to the equator was 51.6 degrees. The launch was thus made in a precise easterly direction.

Part 2 of the flight profile took place outside the radio range of the Soviet Union, over the Gulf of Guinea. There the Block L was automatically ignited by a timing circuit, and it accelerated the combination to the lunar transfer velocity, which resulted in a highly elliptical Earth orbit with an apogee of 310,685 miles. In doing so it aimed at a position in the sky where in three days it would be at the calculated rendezvous point with the moon.

On the way to the moon, the achieved flight path was measured from the control center. During the test phase the probe initiated a roll. It turned about its longitudinal axis once every ninety seconds in order to keep the thermal load low. Then on the second day of the flight, the probe's propulsion unit, the so-called KTDU, began a course correction maneuver of about 164 to 329 feet per second, which placed the spacecraft on the precise course to the intended target area on the moon. This maneuver was carried out by using the probe's thrusters and required a burn time of about fifty seconds.

The landing took place in the fourth phase of the flight. For the time, this was a highly complex undertaking. First the probe's rotation had to be stopped at a distance of 5,157 miles from the moon. Then its position in space was determined with the help of an optical-mechanical system, and the probe was positioned vertical to the surface of the moon. At an altitude of 46.6 miles, exactly defined by the radar altimeter, the KTDU's main engine ignited. Almost simultaneously the probe's side modules were separated and the airbags that were supposed to protect the landing capsule on impact with the surface of the moon were inflated. At about 820 feet above the surface, determined by an accelerometer, the main engine was shut down, leaving the probe's four side-mounted control thrusters still running. At 16 feet above the surface, a sensor made ground contact and the thrusters also shut down, initiating release of the landing capsule. It struck the surface of the moon at a velocity of about 65 feet per second, bounced several times, and then came to rest. The probe's four shell sections then deployed, placing it in an upright position, and at the same time the camera and instruments were released.

Four E-6 landing probes were initially built, with no. 1 being purely a test model not intended for flight use. The first flight unit, series no. 2, was completed in December 1962 and was immediately delivered to Baikonur. Even the greatest optimists in OKB-1 gave the probe a 10 percent probability of success.

The first launch of an E-6 unit took place on January 4, 1963. The first three stages of the R-7 8K78 worked perfectly, but the engine of the Block L, which was supposed to fire high over the coast of West Africa, failed, and the combination was stranded in low Earth

Mission Data, Sputnik 25	
Mission designation	Sputnik 25
Date	January 4, 1963, 09:49 CET
Spacecraft	E-6 no. 2
Booster rocket	8K78L (T103-09) Molniya M
Spacecraft weight	3,300 pounds
Planned mission objective	Soft landing
Mission results	Achieved low Earth orbit only. Insertion into the lunar transfer trajectory failed
Location	Reentered Earth's atmosphere on January 11

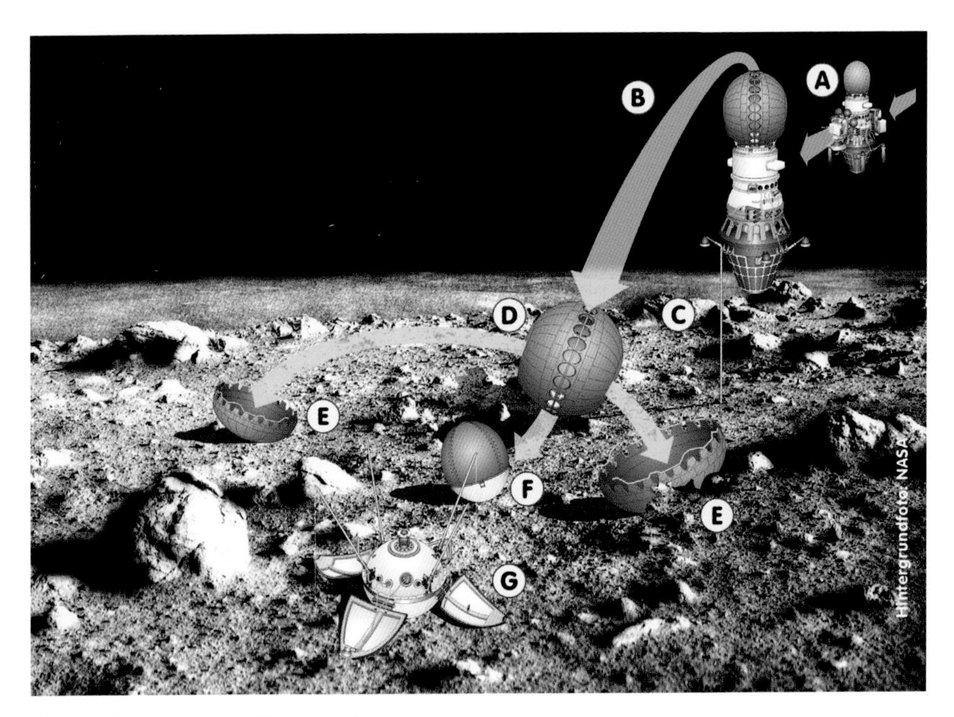

The landing method of lunar probes Luna 4 to Luna 14.

orbit. Including the failures of booster rockets for the Venera and Mars program boosters, this was the sixth case in a row in which the Block L stage stopped working outside the reception range of the Soviet telemetry stations. The Mars and Venus program was plagued by the same problem, since its transfer trajectories also required ignition of the Block L engine in the area of the equator.

This time, however, measures had been taken to obtain data about the cause, should the error repeat itself. The only possible way of learning about the problem was to equip a ship with telemetry reception equipment and send it to the Gulf of Guinea. Prior to the launch of E-6 no. 2, the trawler *Dolinsk* was hastily fitted with the necessary receiver systems, and some of the best engine and telemetry specialists were put aboard for the trip.

This time when the engine of the Block L again failed to ignite, telemetry data were at least assured. Evaluation of the data revealed no clear cause, however. It was suspected that the failure was attributable to a voltage regulator in the newly designed I-100 instrument section. It prevented the command signal from the control system getting through to the Block L engine. The engineers traced this back to the nitrogen atmosphere in the pressure-ventilated container and found that it was too dry, causing the insulation to become brittle and resulting in a short circuit.

The theory was not fully convincing. Andronik Yosifan, head of the All Union Research Institute for Electro-Mechanics, from which the unit came, did not believe this and tested all of the device's units he had available. All lasted much longer than required for a

normal mission. Nevertheless, the nitrogen in the next flight unit was moistened and a small amount of oxygen was also added. It was hoped that this measure would counter the insulation's tendency to become brittle.

The second attempt to launch an E-6 probe was undertaken on February 3, 1963. This time the effort did not reach the ignition of the fourth stage. Telemetry recordings showed unusual behavior by the third stage, which shortly after the beginning of its burn phase entered an elegant arc and continued accelerating toward the Earth instead of heading for orbit. The combination of third stage, fourth stage, and E-6 spacecraft no. 3 subsequently burned up in Earth's atmosphere over Hawaii, watched by the Americans, who had an extensive radar network. During this hot phase of the Cold War they were not particularly amused that an unannounced Soviet projectile should enter Earth's atmosphere right over their beloved Hawaiian Islands. The Western press expressed its disgust that another classic example of Soviet aggression had played out before the eyes of the world.

Mission Data, E-6 No. 3	
Mission designation	—
Date	February 3, 1963, 10:26 CET
Spacecraft	E-6 no. 3
Booster rocket	8K78L (G103-10) Molniya M
Spacecraft weight	3,135 pounds
Planned mission objective	Soft landing
Mission results	The gyroscope that was supposed to regulate the angle of inclination of the third stage failed. The rocket achieved only a suborbital trajectory
Location	After thirty minutes, reentered Earth's atmosphere and burned up over the Pacific

This time the fault was traced to an attitude control gyro. The unit's initial parameters had been entered with too-much leeway relative to the base flight coordinates of the third stage, and the confused third stage headed toward Hawaii, seeking the correct flight path. This gyro system was also in the newly designed I-100 instrument section, and so Korolev's anger was again directed at Nikolai Pilyugin of the NII-885, which was in charge of systems for the I-100.

After this latest failure, Pilyugin's company thoroughly redesigned the E-100 unit. Testing was made more strenuous, and all cables were treated with a paint overcoat. Nevertheless, after seven more failures in a row for the booster, no one suspected that the 8K78 would function perfectly during the next attempt. And yet that is what happened. Ironically it was the thirteenth flight unit of the 8K78 overall. All four stages functioned perfectly, and E-6 no. 4 flew to the moon. In a communiqué, TASS announced to the world: "On April 2, 1963, the Soviet Union launched a space vehicle to the moon. The booster rocket transported the automatic station Luna 4, with a weight of 3,135 pounds. The automatic station Luna 4 will reach the moon in three and a half days."

TASS cautiously avoided mentioning the mission's objective, a soft landing on the moon.

By then the OKB-1 technicians had so little faith in the 8K78 that they were relatively surprised by the successful launch. An initial trajectory analysis revealed a rather precise trajectory insertion, and all the specialists made their way to the tracking station in Crimea. But despite all the efforts and calculations to determine the trajectory, on April 6, the probe inexplicably flew past the moon at a distance of 5,280 miles.

The landing unit of the moon probes Luna 4 to Luna 14.

The now-obligatory investigating committee found several minor faults and one significant one, and it was in the probe's astronavigation system, the so-called SAN. This device was also integrated into the delicate I-100 unit. It turned out that that reacted very sensitively to temperature variations. Once again, improvements were made to the system as well as several more redundancy components, which improved the accuracy of orbit determination. The weakest part of the system remained the 8K78; the Venera and Mars program was also plagued by bizarre errors, most of which had one thing in common: the failure of the Block L stage.

Mission Data, Luna 4	
Mission designation	Luna 4
Date	April 2, 1963, 09:04 CET
Spacecraft	E-6 no. 4
Booster rocket	8K78L (G103-11) Molniya M
Spacecraft weight	3,300 pounds
Planned mission objective	Soft landing
Mission results	Insertion into transfer trajectory successful. Course correction failed. Flew past the moon at a distance of 5,182 miles
Last contact	April 6, 1963
Location	Initially in Earth orbit with a perigee of 55,800 miles and an apogee of 434,028 miles. Later entered a heliocentric orbit

May 15: Gordon Cooper circles the Earth.

After the double mission of Vostok 3 and 4, the Soviets were relatively certain that the next two missions would follow this scheme. Once again, the principal idea was to demonstrate Soviet superiority in space to the

Valentina Tereshkova

Americans. A heated discussion developed as to whether the new double mission should be carried out by two women, one man and one woman, or two men. The single constant in the dispute, which continued in these weeks, was that at least one of the two spacecraft should surpass the flight record of Vostok 3 and remain in orbit for up to eight days.

The idea of training female cosmonauts appeared soon after Gagarin's flight. Then in February 1962, five women were selected from 400 candidates: Valentina Ponomaryova, Irina Solovyova, Sanna Yerkina, Tatyana Kuznetsova, and Valentina Tereshkova. All were between twenty and twenty-eight years old.

Korolev was very much opposed to flights by women at this early stage of the manned spaceflight program. In his opinion, further experience should be gained first, and the missions should gradually be extended to six and later to eight days. Above all, the coming flights were to see scientific work done onboard the spacecraft.

Kaymanin, on the other hand, fought to have at least one woman on both Vostok 5 and Vostok 6. In doing so he followed the party line that every effort should be made to demonstrate the superiority of the Soviet system in the field of space travel and to demonstrate to the world that in socialism, women played an absolutely equal role to men. The final decision was made in April 1963. The planned double mission would see one man and one woman go into space.

There were no controversies when it came to the male cosmonauts. Kaymanin chose Valeri Bykovsky to be pilot, and his alternate was Boris Volynov, and everyone agreed on their selection. The problem was the woman who was to fly Vostok 6. Keldysh and several other officials, such as Marshall Rudenko, who was close to the Academy of Sciences, favored Valentina Ponomaryova, who had an academic background. She proposed Valentina Tereshkova as her "alternate female." Gagarin and Korolev, along with several others, favored Tereshkova as primary pilot and Ponomaryova as her alternate. The reason for this was not particularly objective. While Tereshkova was not particularly experienced and was less trained, she came from a working family, was active in the party, and in particular was better looking than the rather austere

Ponomaryova. She was thus more marketable as a "token socialist woman." Valentina Ponomaryova, who was a trained engineer, would surely have been better qualified than Valentina Tereshkova. She was independent, self-confident, and probably even better qualified for the mission than some of her male colleagues, but the standards of the time could not accept that.

In May, the engineers and managers reported the spacecraft and rockets ready to go, but they complained that they still did not know for whom the contoured seats should be made. Finally all of the decision makers got together and agreed that Tereshkova should fly. Her flight duration was not to exceed three days, while an attempt was to be made to keep Vostok 5 and Bykovsky in orbit for up to eight days.

After some delays, initially because of concerns about increased solar radiation, and then because of very windy weather (wind speed near the ground had exceeded the allowable limits for the R-7), Bykovsky finally blasted off on June 14. There were no problems during launch, but the rocket placed Vostok 5 into a worryingly low orbit. The perigee, or the lowest point in the orbit, was just 100 miles, while the apogee, the highest point, was barely 130 miles. Since the Vostok spacecraft was incapable of making any changes in orbit, the flight controllers were forced to live with the orbit into which the booster had placed it. Initial calculations showed that this orbit would permit a flight of only five days, and on June 17 the decision was made to return Bykovsky to Earth after five days in space and eighty-two orbits of the Earth.

The launch of the twenty-six-year-old Tereshkova on June 16 was also trouble free. It took place at 10:29 Central European Time. The achieved orbital parameters were a perigee of 112 miles, an apogee of 144 miles, and an orbital inclination to the equator of 64.95 degrees. Her flight program foresaw a landing after the forty-ninth orbit. The flight controllers on the ground soon began to have doubts about her capabilities, since she began giving many unclear and deviant responses, and several times she forgot or intentionally failed to activate radio communications. It was later revealed that she had been battling space sickness but failed to admit it in radio communications with ground controllers.

In her later reports she expressed criticism about several of the more trivial aspects of the mission, things that Bykovsky did not even mention. She regretted not having had the ability to brush her teeth and complained about the dryness of the bread that was part of her rations. She also complained that her space suit became increasingly uncomfortable as time went on. Things that would have been considered design criticisms from a male cosmonaut were later attributed to her "bitchiness."

Initially, however, the attention of the controllers fell upon Bykovsky; on June 18, shortly after nine in the morning, he reported a "knock" (*stuk* in Russian) in space. This caused some concern, especially because the quality of the radio transmission was terrible. The control center hastily devised a list of questions about this mysterious occurrence to ask Bykovsky during the next radio contact.

At the very start of the contact about ninety minutes later, they asked Bykovsky about the nature and origin of the "knock." He, however, had no idea as to what the controllers were talking about. They had to read out the text of the radio report to him once again, then it came to him. The controllers had misunderstood him. He had not said "*stuk*" but "*stul*," meaning that he had had a bowel movement.

A more serious concern had meanwhile arisen as to whether Tereshkova would be capable of manually positioning the spacecraft for reentry. On the first day of the mission it had been her task to manually carry out the attitude adjustment of the spacecraft. This was a sort of practical test for the scenario of failure of the automatic system. She failed and subsequently received "help" by radio.

On June 19, at about 13:00 Central European Time, the activation command for an automatic landing was transmitted by radio to both spacecraft almost simultaneously. Bykovsky gave a running commentary on the progress of his return, while Tereshkova stayed silent. Either something was wrong with her communications equipment or she intentionally did not report, or perhaps something else was wrong. No one knew. Communications were also a considerable problem area in Soviet space travel, however. Not until about 15:00, one and a half hours after the estimated landing time, did it become clear that all was well with both cosmonauts.

In the end there was another mission. A correspondent and several residents of the landing zone arrived on the scene before the official physicians, officers, and technicians who were in fact responsible for Tereshkova's safety. Some had brought food, and they spread it out on Tereshkova's parachute. Then all of them, the female cosmonaut and her visitors, ate their fill of the delicacies. Tereshkova shared her "cosmonaut rations" with her "guests," and in the end it was impossible to determine how much she had actually consumed onboard the spacecraft. They were therefore unable to verify her contradictory statements about the condition of her rations and her appetite.

After the mission the discussion about whether women were suitable space travelers flared up again. It ended in the conclusion that trained pilots were obviously necessary in order to master critical situations during a space flight. While Tereshkova had been a parachutist, she was no pilot. In addition, the women on the team had received only abbreviated training compared to their male counterparts. That was also identified as an error.

Korolev did not apologize, however. He was quite dissatisfied with Tereshkova's performance and in particular by her undisciplined behavior in omitting reports from space and consuming her rations after landing. His statement is still quoted to the present day: "As long as it is up to me, there will be no more girls in space." But whatever. The result was that Tereshkova alone had spent twice as much time in space as all the American astronauts put together.

Valentina Tereshkova and Yuri Gagarin.

1964: FIRST HALF OF THE YEAR

January 30: The mission of Ranger 6 was another failure. While the space probe precisely reached its target and transmitted telemetry until the end, its six cameras sent no pictures. This was attributed to a short circuit that had occurred during the early launch phase of the booster rocket, during which the camera system was switched on and off again for about a minute.

The fourth attempt in the E-6 program to place a probe on the moon took place on March 21, 1964. It was the one hundredth flight by a Zemyorka from launch site 1/5 at Baikonur. The mission ended very quickly. This time Block I, the third stage of the R-7, was the guilty party. The main oxidizer valve only partially opened because a safety rod had broken off, and the engine shut down in the 489th second of flight.

A mission to Venus was scheduled for March 27, just a few days later. This time the Block L stage failed and the combination of spacecraft and stage remained in a low Earth orbit. Onboard the improved 8K78, which was now called the 8K78M (or also Molniya M), was a data recorder, from which data could be retrieved as it passed over radio stations. It worked, however, only if the rocket reached at least a parking orbit around Earth. It revealed that there was a fundamental design flaw in the wiring of the power supply. Because of the lack of power, the attitude control gyroscope did not run up, the Block L stage with the Venera spacecraft tumbled in its orbit, and the engine failed to start because of the absence of flight attitude information. It was a problem that a technician on the ground could fix in twenty minutes with a simple bypass. An entire series of failed starts by the Block L stage was due to this trivial cause.

The fifth attempt to land on the moon began on April 20, 1964. Block L did not receive an ignition command from the I-100 unit, and that was the end for E-6 no. 6. The rocket did not even reach orbit, crashing into

Mission Data, E-6 No. 6	
Mission designation	—
Date	March 21, 1964, 09:16 CET
Spacecraft	E-6 no. 6
Booster rocket	8K78M (T15000-20) Molniya M
Spacecraft weight	3,135 pounds
Planned mission objective	Soft landing
Mission results	The rocket failed to reach Earth orbit because of a defective fuel valve in the third stage
Fate	Reentered Earth's atmosphere after about 30 minutes

Mission Data, E-6 No. 5	
Mission designation	—
Date	April 20, 1964, 09:28 CET
Spacecraft	E-6 no. 5
Booster rocket	8K78M (T15000-21) Molniya M
Spacecraft weight	3,135 pounds
Planned mission objective	Soft landing
Mission results	The booster rocket's flight control system failed 340 seconds after liftoff
Fate	Reentered Earth's atmosphere over Siberia

the Pacific Ocean. The *Dolinsk* was again on tracking duty and two weeks later returned with telemetry tapes. Despite the considerable experience with this equally complex and sensitive unit, it was obvious that the flight controllers still did not have sufficient knowledge of its behavior in space. It was found that there was considerable overheating in various places in the interior. The causes of this problem remained a mystery, however. Once again a heat-damaged power cable was the suspected source, and so it was proposed that the unit be cooled prior to launch. This suggestion found no enthusiastic supporters, since special treatment of the unit, buried deep inside the rocket until just before launch, would significantly complicate the launch preparations.

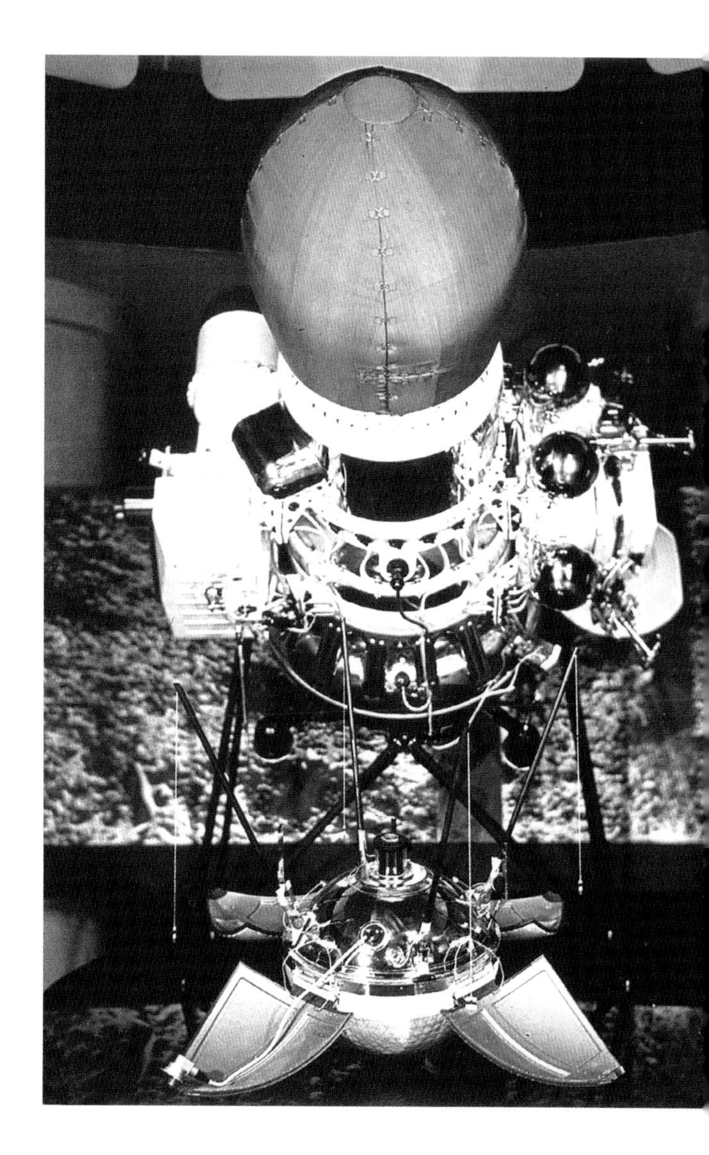

Model of the space probe Luna 9. At the top of the photo, the landing capsule is in flight configuration; in the lower part of the photo, it is in opened configuration after landing.

1964: SECOND HALF OF THE YEAR

At the end of July 1964, a government decree authorized the development of Korolev's L3 moon lander complex. It also authorized a research program for the development of engines and infrastructure for liquid hydrogen–liquid oxygen fuel. Korolev regarded this as absolutely necessary in order to improve the future performance of the N1 so that a manned moon landing could be carried out with a single booster rocket.

July 28: Ranger 7 becomes the first successful American lunar probe. In the seventeen minutes prior to impact in the Mare Cognitum, the spacecraft sends 4,300 photographs to Earth.

On August 3, 1964, the Soviet government issued another decree. Its subject was the development of a circumlunar spacecraft for a manned flight in the second half of 1967. This mission, however, was to use Vladimir Chelomei's LK-1 as the spacecraft and its UR-500K booster rocket, better known by the name Proton.

By 1963, it was already clear that the new Soyuz spacecraft would not be ready to fly in 1964. From the Americans were coming unsettling reports about the state of the Gemini and Apollo programs. The three-seat Apollo was supposed to make its first manned flight in 1966, and the two-seat Gemini at the end of 1964. These spacecraft were also far more capable than the Soviet Union's Vostok system. It was felt that the period until the Soyuz was ready, which was expected to be toward the end of 1965, had to be bridged with further space firsts in order to maintain the nation's lead in space over the US.

To Korolev's great displeasure, at the turn of 1963–64, Khruschev ordered the Vostok modified to accommodate three people. Khruschev had of course not come up with this idea himself, even though he was very interested in the space program and was unusually well informed for a leading politician. Instead it was his son Sergey, who was a spaceflight engineer in Vladimir Chelomei's OKB-52 and who instructed his father accordingly. The government decree that made the matter official was not signed until June 14, 1964. By that time, work was fully under way.

At first it seemed completely hopeless to redesign the cockpit of the Vostok, which was laid out for a single cosmonaut, to accommodate three. Korolev himself also thought it impossible, and it required a number of brilliant ideas and great courage to take risks to ultimately make it possible. The modifications were given the name Voshkod (sunrise), which Western observers assumed meant that it was a completely new vehicle. In fact it was nothing other than a Vostok capsule that had been so emptied of components that with great difficulty and with a special seating arrangement, it could accommodate three people. These three cosmonauts could fit into the cramped ship only while wearing training suits. Many of the safety precautions for launch and recovery also had to be eliminated. There were no ejection seats, and during the launch phase the crew had to rely on the rocket's reliability for better or for worse. The capsule could not have accommodated two ejection seats, to say nothing of three. Thus, the landing proce-

Integration of Voshkod 2. The inflatable airlock can be seen clearly. In the background is the 11A57 special version of the R-7 for Voshkod launches.

dure also had to be changed so that all three space travelers could be returned to the Earth together in the capsule. The single cosmonaut in the Vostok exited the capsule by ejection seat prior to landing, and the cosmonaut and capsule landed separately. This was no longer possible with the Voshkod.

The supplies onboard were now sufficient for just two days. If there was an error in the

The R-7 11A57 booster rocket with Voshkod 2 immediately prior to launch.

orbital insertion combined with a failure of the retrorocket system, the crew was lost. The Vostok carried so many supplies that it was possible to wait for the braking effect of the upper atmosphere to return the cosmonaut to Earth.

Despite all the weight-saving measures, in the end the Voshkod weighed almost 2,200 pounds more than the Vostok. This payload could no longer be placed in orbit by the 8A72 version of the R-7 previously used. They now needed the 11A57 variant, whose third stage used Semyon Kosberg's RD-108 Block L engine, producing just 67,442 pounds of thrust.

As had been the case with Vostok 6, a bitter dispute now broke out over the composition of the crew. From the beginning it was clear that only one of the three crew members had to be an experienced cosmonaut. This man from Kamanin's cosmonaut team was to act as commander, while the other two members of the crew were supposed to be a doctor and an engineer. Kamanin insisted, however, that these two others must also come from the air force. The participation of civilians in space flights threatened his jealously guarded monopoly.

Kamanin was unable to get his way, however. The Health Ministry wanted to contribute the doctor and named Boris Yegorov. And Korolev was determined that his star designer, Konstantin Feokistov, should be part of the crew. He had been responsible for the Zenit and Vostok systems and knew the Voshkod inside and out, much better than any cosmonaut.

Kamanin finally accepted Yegorov, but having a second civilian, especially one of Korolev's closest associates, in the spacecraft was a real problem for him. Additionally, Feokistov did not meet the medical requirements for a cosmonaut. He was relatively large, was too old by the standards of the day, and to top it off was also nearsighted. Korolev fought with the gloves off and threatened Kamanin that he would form his own corps of cosmonauts. His argument: "Engineers can fly this space vehicle just as well." Kamanin finally accepted.

An unmanned test mission by the complete Voshkod system took place on October 6, and was called Cosmos 47. Its primary purpose was to test the telemetry, life-support system, and new landing procedure, with a braking rocket in the lower part of the parachute system. The mission was largely error free, and on October 12, 1964, the crew under the command of Vladimir Komarov launched on a twenty-four-hour mission.

Somewhat unexpectedly and in contrast to their superiors, the three men got along famously, and from a technical standpoint the mission was trouble free. After twenty-four hours and seventeen minutes, Voshkod 1 landed in Kazakhstan. It was the first manned landing on solid ground in which the crew remained in the spacecraft.

As hoped, the West again reacted with shock. The three-man spacecraft's design and performance were falsely compared with those of the three-man Apollo spacecraft, and it was suspected that the Soviets were still more than two years ahead of the West in the race to the moon.

1965: FIRST HALF OF THE YEAR

February 17: Ranger 8 is a complete success. In the twenty-three minutes prior to impact in the Mare Tranquillitatis, it sends 7,137 photographs to Earth.

On March 12, 1965, the Soviets launched E-6 no. 6. This time, while it achieved a parking orbit, Block L again failed to ignite for entry into the lunar transfer trajectory, and it remained in an orbit with a perigee of 125 miles and an apogee of 178 miles and was given the unassuming name Cosmos 60. Once again the telemetry data directed suspicion for the failure to the suspect I-100 unit. Once more it was examined closely, and this time, after many hours of testing with another unit, a slightly projecting washer was discovered on one of the shafts of the PT-500 transformer unit, which under certain thermal conditions rubbed against a screw head of the case cover and could cause the shaft to fail. Whether this was perhaps far-fetched and whether it had in fact taken place during the flight, no one could say.

In any case, it was decided to replace the entire PT-500, a task that should actually have required the entire Block L back to the factory where it had been made. This, however, was regarded as an unacceptable waste of time, since it would have taken many weeks. And so, a way was found to undertake the modification at Baikonur. Instead of the PT-500 converter, two PT-200 units were installed. This was not particularly complex, but all the wiring had to be changed, the test procedures now ran differently, and the thermal characteristics of the entire I-100 unit changed again.

The Soviets hoped to achieve another first in manned spaceflight before the Americans began the manned flights of the Gemini program: the first exit of a spacecraft by a man in space. The Voshkod had to be modified significantly in preparation for this mission. Soviet engineers designed an inflatable airlock

Mission Data, Cosmos 60	
Mission designation	Cosmos 60
Date	March 12, 1965, 09:28 CET
Spacecraft	E-6 no. 9
Booster rocket	8K78M (T15000-21) Molniya
Spacecraft weight	3,500 pounds
Planned mission objective	Soft landing
Mission results	Low Earth orbit (125 x 178 miles). Ignition of the third stage for the translunar insertion maneuver failed
Fate	Reentered Earth's atmosphere on March 17, 1965

Voshkod 2 primary crew: left, *Alexey Leonov;* right, *Pavel Belyayev.*

Voshkod 2 backup crew: left, *Yevgeny Khrunov;* right, *Dimitri Zaikin.*

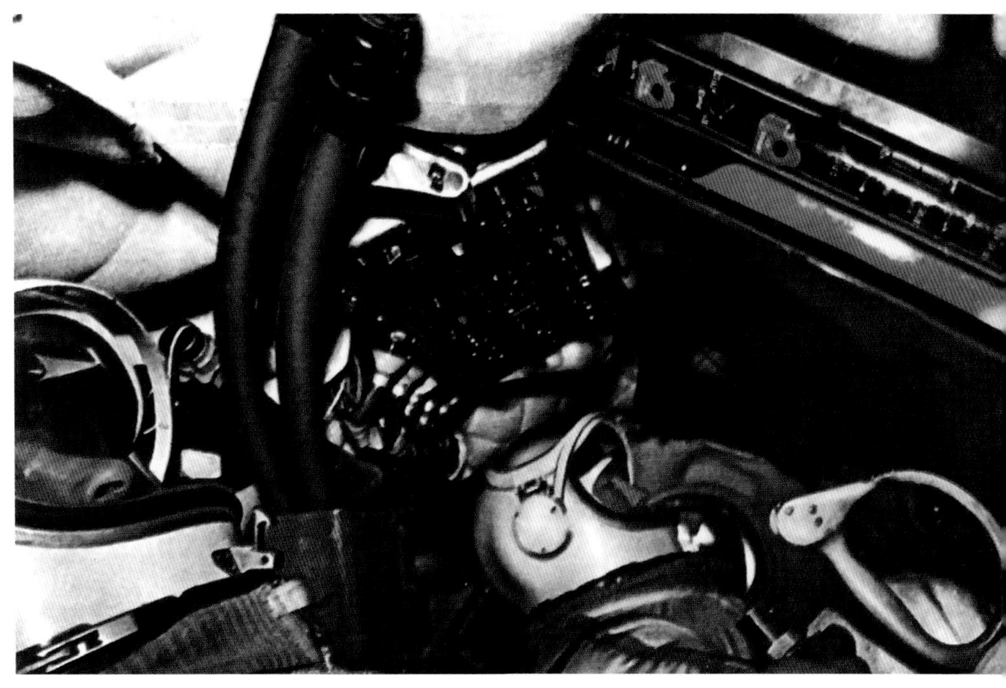

Leonov and Belyayev during simulator training for the Voshkod 2 mission.

for the spacewalk. Through it the cosmonaut could leave the spacecraft and after his excursion in space return to the cabin again. Once the maneuver had been completed, the airlock was to be jettisoned. Although there were only two cosmonauts onboard this time, because of the 550-pound airlock it was again impossible to take supplies for more than two days. Korolev had promised Khruschev (before he was deposed by Brezhnev and Kosygin) that he would undertake preparations for this mission. It was thus contained in the decree of June 1964. Like the previous Voshkod 1 mission, an unmanned orbital test mission was planned before the manned flight took place. Named Cosmos 57, it was launched on February 22. The test went perfectly, at least in the beginning. The airlock inflated as planned, and good-quality images were sent to Earth by the newly installed TV transmission system. After the third orbit, however, contact with Voshkod 2 was lost and it disappeared without a trace.

As usual, an investigating committee was formed and soon afterward it discovered the embarrassing result. Two ground stations in Kamchatka had, independently of one another, twice sent the control command no. 42 for the airlock to the spacecraft. However the decoder onboard mistook the two signals arriving almost simultaneously as signal no. 5. This was the command that was supposed to initiate the spacecraft's descent into Earth's atmosphere. Cosmos 57's control logic correctly noted that the spacecraft would land outside the territory of the USSR, and activated the

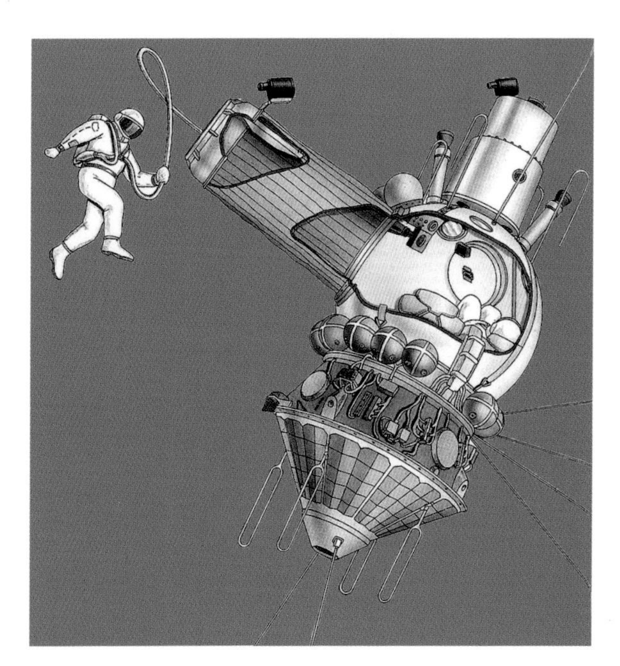

Diagram of Leonov's spacewalk.

vessel's self-destruct mechanism. In the crazy days of the Cold War this had been installed to prevent the secret vehicle from landing in foreign territory and falling into the hands of the "class enemy." One must rightfully say that only the unmanned space vehicles of the Soviet Union had such a mechanism. The manned spacecraft were understandably not equipped with it.

On the evening before the launch of Voshkod 2, the primary crew was confirmed by the state committee. It consisted of commander Pavel Belyayev and pilot Alexey Leonov. It was the latter who was to carry out the exit maneuver. The backup crew consisted of commander Dimitri Zaikin and EVA cosmonaut Yevgeny Khrunov.

On March 18, 1965, just one week before the first manned flight of the American Gemini program, Belyayev and Leonov launched into Earth orbit. Just ninety minutes after launch, Leonov left the spacecraft through the airlock and for twenty minutes floated next to the Voshkod. Then he climbed back into the vehicle, and after twenty-four hours in orbit the two men landed safely on Earth.

This brief description was how the Soviet media presented the mission of Voshkod 2 to the West—a technically perfect tour de force that had gone entirely according to plan. It was not until decades later that it was learned that the mission had been a disaster from start to finish, and that the cosmonauts had been lucky to survive.

The airlock of Voshkod 2 had the advantage that—unlike the later Gemini space capsules—the spacecraft's atmosphere did not have to be completely evacuated when the cosmonaut exited. The other cosmonaut, in this case Belyayev, could remain in his pressurized cabin. On the other hand he was unable to come to the aid of his comrade should he run into difficulties. And exactly that happened to Leonov. During the twelve-

minute spacewalk, his spacesuit became so inflated that on his return he could no longer fit into the airlock, which was only 28 inches wide. After several unsuccessful attempts, faced with the possibility of not being able to regain the safety of the spacecraft, he reduced the pressure in his spacesuit to a dangerously low level. He squeezed himself in headfirst, instead of feet first as planned. With great effort he managed to find space in the airlock again.

After Leonov had returned to the cabin, the airlock was jettisoned. This resulted in a drop in pressure in the containers for the cabin air. It fell from 75 to 25 atmospheres. The return to Earth would have to take place by the seventeenth orbit at the latest; otherwise the crew would run out of air. Flight control decided to break off the mission during the sixteenth orbit of Earth. That was the first time that Voshkod 2 passed over the planned landing area.

But then the automatic-orientation system failed, whereby as a result the retrorocket did not ignite. In great haste, preparations were made so that Belyayev could manually initiate the return to Earth at the end of the eighteenth orbit. Afterward there were no radio communications with the cosmonauts for four hours. Finally, from a search helicopter came the report that instead of landing in the Kazakh steppe, the capsule had come down 1,200 miles from the planned

Another photo of the Voshkod 2 booster rocket.

Alexey Leonov during his extravehicular activity.

landing point. Its location was 18 miles southeast of the city of Berezniki in the Urals, in a dense forest covered in deep snow. Belyayev had initiated the manual recovery forty-eight seconds too late, which caused the capsule to overshoot the landing zone.

The two cosmonauts survived the rough landing with contusions and bruises, and they were able to indicate their presence by using Morse code. Nevertheless it was four hours before a helicopter spotted them. The dense forest made it impossible for the helicopter to land, and the difficult terrain even made a recovery with the cable winch impossible. Finally it became night and the action had to be called off. Recovery forces did not reach the pair until twenty-four hours after their crash landing in the forest. The cosmonauts then had to bivouac in the snowy forest with them for another night. The next morning they made their way on skis to a clearing, 1.2 miles from the landing site. Special forces had cut down trees and bushes there to create a landing site for the rescue helicopter.

Although in the end the mission became one last triumph for manned Soviet spaceflight, even though no one realized it at the time, Voshkod 2 was the turning point in the space race between the Soviet Union and the United States. From then on the US took the lead, but in that March of 1965, it appeared that the Soviet Union had once again demonstrated its superiority in space.

March 21: Ranger 9 mission is a complete success. In the final nineteen minutes prior to impact in the Alphonsus crater, it sent 5,814 photographs back to Earth.

March 23: Gemini 3, the first manned Gemini spacecraft, was launched with astronauts Grissom and Young on a flight lasting four hours and fifty-three minutes. The spacecraft circled the Earth three times.

On April 10, 1965, the seventh E-7, production unit no. 8, was launched, and a few minutes after it had left the launchpad it was clear that another accident-investigating committee would be required. The precise cause of the crash could not be determined. The committee suspected, however, that the operating pressure of the Block L third-stage engine did not build up as expected because of a leak in the tank pressure regulation system. The oxidizer and fuel tanks failed to reach the necessary operating pressure. Another possibility, which was considered less likely, was that the engine's combustion chamber had burned through. Nothing precise could be found, since the "lunar train," as the Russian engineers called the combination of Block I and the lunar vehicle, sank in the Pacific.

Mission Data, E-6 No. 8	
Mission designation	—
Date	March 12, 1965, 09:28 CET
Spacecraft	E-6 no. 8
Booster rocket	8K78L (R-103-26) Molniya
Spacecraft weight	3,135 pounds
Planned mission objective	Soft landing
Mission results	A nitrogen pressure tank in the third stage failed, which caused a pressure loss in the oxygen tank, cutting off the flow of oxidizer and finally leading to shutdown of the engine
Fate	Reentered Earth's atmosphere over the northern Pacific

The next attempt to launch an E-6 probe was scheduled for May 9 because of reasons of celestial mechanics, even though that was one of the Soviet Union's "holiest" holidays (the day of the victory over Germany). The launch of a Molniya 1 satellite was, however, planned for April 23, between the launch of E-6 no. 8 on April 10 and the launch of E-6 no. 10. The Molniya were also launched using the 8K78. The version used for these launches had a significantly lower failure rate because it was not equipped with the highly complex and extremely error-prone I-100 system. The launch of the Molniya on April 23 was a success, the Block L stage functioned reliably, and the satellite was ready for service on May 9.

This seemed to be a good omen, since the launch of the E-6 with the serial no. 10 was a success. The tracking vessels *Krasnodar* in the Mediterranean and *Dolinsk* in the Gulf of Guinea transmitted radio messages via short wave that the Block L stage had functioned well.

Day and night, tracking station NIP-10 in Crimea compared the flight parameters of the space probe (which had meanwhile been officially named Luna 5) with the planned values. From time to time there were problems with the radio link, since a series of operational problems with the tracking system interfered with communications with the probe.

During one of these communication sessions, the link became worse from minute to minute. It was suspected that the interference came from the radio systems of the Black Sea Fleet and the air defense system in Crimea. Korolev immediately contacted the commander of the fleet and asked him to turn off all radio systems. The admiral stated that he was

The legendary tracking vessel Dolinsk, *always camouflaged as a trawler in the 1960s, here seen in August 1983. Only a shadow of its former glory.*

prepared to halt all ship-to-ship communications for a time but could not promise the same for the air defense zone. Just as Korolev began negotiating with the responsible air defense posts, the mysterious problems with radio traffic miraculously cleared up. The leaders of the station had neglected to provide the maser, the receiving part of the 105-foot-diameter antenna, with liquid nitrogen coolant. It had become increasingly warm, causing an increase in static noise. Korolev, whose outbursts of rage over such things had become legendary, accepted the explanation with a gentle smile, returned to the telephone, and apologized to the admiral and the responsible air defense units.

But in the end, this did not help Luna 5. E-6 no. 10 lost its proper position in space when the first attempt was made to carry out a corrective maneuver. With some difficulty the controllers managed to stabilize the space probe again. With haste they tried to analyze what had caused the probe's somersaults. They quickly discovered that the warming-up phase for the attitude control gyro had been too short. When carrying out the corrective maneuver with the main engine, they failed to achieve the necessary change in attitude.

During the second attempt the inexperienced control team made an error, wrongly calculating the attitude control values. For the whole three days of the flight to the moon, the flight controllers struggled to at least land the probe somewhere on the moon, even if not in the planned target area. In any case they succeeded, and the probe struck the moon. It did not make a soft landing, however, again because of the gyroscope, which failed to gain sufficient speed. The probe finally landed hard on the moon, more than 400 miles from the planned landing site. The loss was a painful one, but there was a certain amount of pride. The engineers and technicians of the E-6 program had never achieved this before.

For the first time in the history of the Soviet space program, TASS openly admitted a failure in space. Its report read: "Luna 5 reached the surface of the moon near the Sea of Clouds. During the flight to and the approach to the moon, a great deal of data were received, which is of essential importance to further development of a system for a soft landing on the moon."

The thought behind this change of heart was simple. Somehow they had to tell the world why they had sent probe after probe toward the moon. And as well, the Americans had also not succeeded in making a soft landing on the moon.

When all the data were analyzed after the mission, it was found that the orientation of the probe in the decisive moment before the braking engine was ignited had also not succeeded because the pressure-ventilated container for the spacecraft's airbags began shaking so badly that from the beginning it could not maintain its correct position in space.

Mission Data, Luna 5	
Mission designation	Luna 5
Date	May 9, 1965, 08:50 CET
Spacecraft	E-6 no. 10
Booster rocket	8K78M (U-103-30) Molniya M
Spacecraft weight	3,250 pounds
Planned mission objective	Soft landing
Mission results	Landing engine failed to ignite because of a flight control error. Crashed onto the moon
Fate	Impacted at 8 degrees north and 23 degrees west (near Copernicus crater)

June 3: James McDivitt and Edward White launch on the four-day Gemini 4 mission. During the flight, White makes the first American spacewalk.

Just a month later, on June 8, 1965, E-6 no. 9 was sent on its way to the moon. The launch and insertion into the lunar transfer trajectory were faultless, and soon after the launch the Soviet news agency TASS made the official announcement and revealed that the probe was called Luna 6. As they had for the previous flights, the technical heads of OKB-1 again gathered at NIP-10 in Simferopol in Crimea. Everything went perfectly. Too perfectly, as it would soon become clear.

The first eleven communications contacts with the probe were completed successfully, including the first trajectory corrections. The decisive flight path change was envisaged for contact no. 12. It was supposed to put the probe precisely into the planned landing area. But suddenly, to the dismay of mission control, the incoming telemetry data showed that the engine had failed to shut down. It ran until all the fuel onboard had been consumed, including the fuel for the landing itself. The ballisticians calculated that the probe would miss the moon by no less than 99,420 miles.

The problem, which was a serious operational oversight, was discovered within a few minutes. While the code for the execution of the burn maneuver that flight control sent to the probe defined the start time of the burn maneuver, it failed to define its end. The probe understood this to mean that it should fire until automatic shutdown, when its fuel was used up, and that is what it did.

Understandably, at first there was a funereal atmosphere in the control center, but then they made the best of what had happened. They used the off-course but perfectly functioning probe to test all the system's functions up to the landing itself. The radio system was tested to a distance of 360,000 miles, as was the functioning of the redesigned attitude control system, separation of the landing segment from the engine section, and pressurization of the airbags for the shock absorber system.

Mission Data, Luna 6	
Mission designation	Luna 6
Date	June 8, 1965, 08:42 CET
Spacecraft	E-6 no. 7
Booster rocket	8K78M (U-103-31) Molniya M
Spacecraft weight	3,175 pounds
Planned mission objective	Soft landing
Mission results	During the primary course correction on June 9, the engine ran until the fuel was consumed
Fate	Passed the moon at a distance of almost 99,420 miles on June 11. Since then it has been in a heliocentric orbit

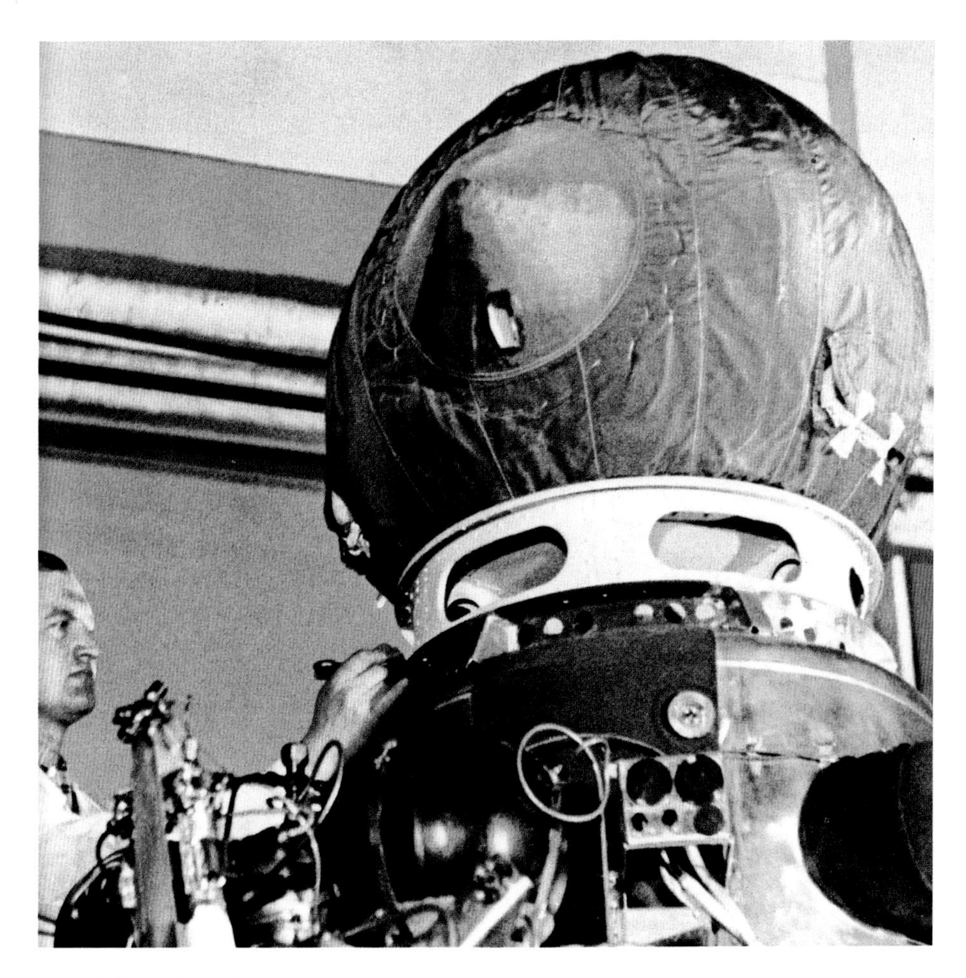

Luna 9 during launch preparations.

1965: SECOND HALF OF THE YEAR

On July 16, 1965, the UR-500 made its successful first flight from the new Launchpad 81 at Baikonur. Its payload was the 13.45-ton research satellite Proton 1. The mission was another propaganda success for the Soviet Union, and it was rolled out by the press. Of the three other test flights, each with another Proton satellite onboard, two were successful. For a completely new rocket, this was a good success by the standards of the day. The UR-500, now called the Proton, was to play a significant, though not always glorious, role in the Soviet Union's circumlunar program.

The next 8K78 with a Block L and an E-6 lunar probe was delivered to Baikonur in August and readied for launch. It was the eleventh probe and the tenth attempt to make

a soft landing on Earth's satellite. The moon shot was planned for September 4, 1965. On that day, the rocket was already fueled and ready to launch when an error in the acceleration sensor was discovered. It was not possible to remedy the fault on the launchpad. The 8K78 therefore had to be defueled and brought back to the integration building. That would have been a matter of several days, but it would mean missing the launch window in the month of September. Time to make an unmanned soft landing before the Americans was slowly running out. Their Surveyor project was already in the starting blocks.

As a precautionary measure and to at least secure another first, Korolev had obtained permission to modify one of the Mars program probes and convert it into a lunar flypast probe. This spacecraft, designated Zond 3, was launched successfully by an 8K78 on July 18, 1965, and four days later it passed the moon at a distance of 5,716 miles. The equipment developed for a photographic survey of Mars also functioned perfectly at the moon, and it sent to Earth photographs and spectral images that were considerably better than those of Luna 3 five years earlier. Even though the Ranger space probes had already sent back extremely sharp close-up photographs of the front side of the moon, these were still the only ones of the back side of the moon.

Zond 3 was essentially a Mars probe. It was hastily modified to obtain good photographs of the dark side of the moon during flypast, ahead of the Americans.

Mission Data, Zond 3	
Mission designation	Zond 3
Date	July 18, 1965
Spacecraft	3MV-4 no. 3
Booster rocket	8K78 Molniya
Spacecraft weight	2,116 pounds
Planned mission objective	Technology demonstrator for deep space missions
Mission results	Took 28 photographs of the moon during flypast at distance of 5,717 miles and transmitted these photos from a distance of 19 million miles from Earth. Remained in operation for 228 days
Last contact	March 1966
Fate	Heliocentric orbit with a perihelion of 0.9 AE and an aphelion of 1.56 AE

August 29: The crew of Gemini 5 (Gordon Cooper and Charles Conrad) carried out an orbital flight lasting eight days and circled the Earth 120 times.

The E-6 probes should long since have made soft landings on the moon by the end of 1965. No. 10 was the last of the first series of probes. In around-the-clock shift work, OKB-1 had produced two more probes, and a third came from Lavochkin. The two OKB-1 probes were given the serial nos. 11 and 12. The probe supplied by Lavochkin, however, was given the number 202, not least because of the superstition associated with the number 13. No. 201 had been the test device built at the beginning of the series. It was never used in space and was used only for ground trials to test modifications.

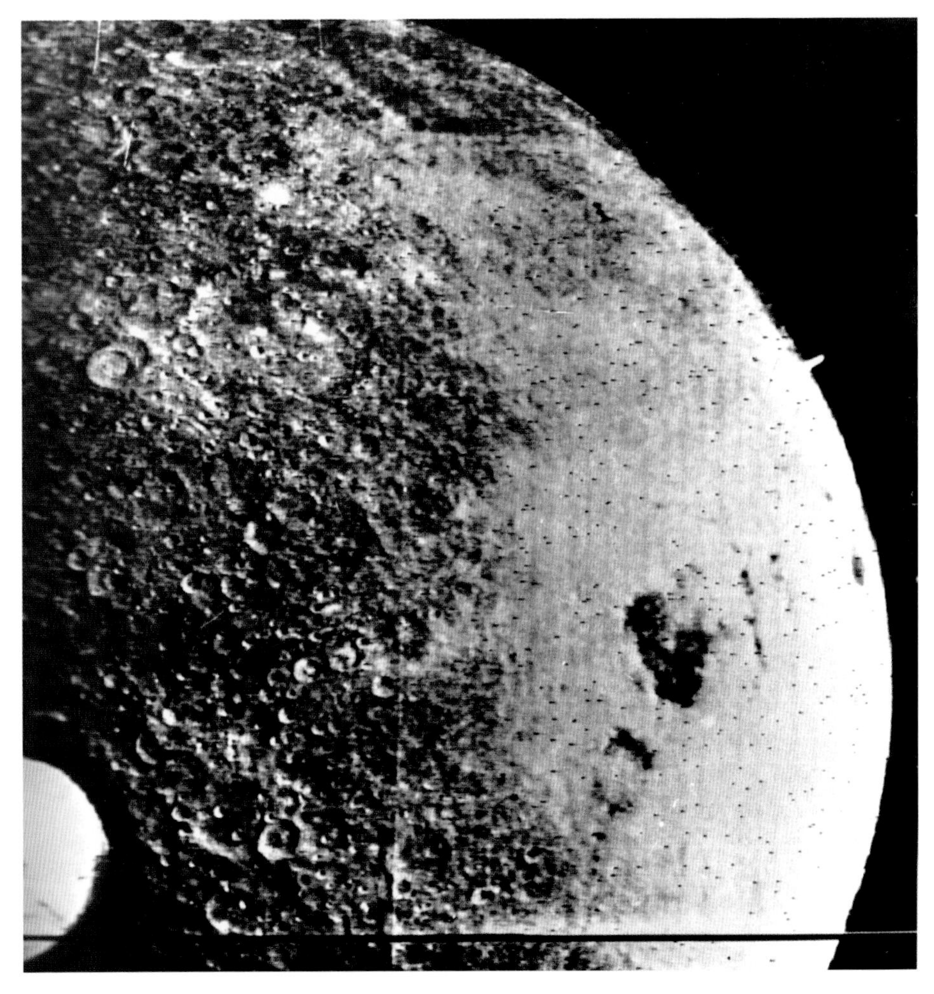

One of the pictures taken by Zond 3. The Tsiolkovsky crater can be seen in the right center of the photo.

The launch date for landing probe no. 11 was set for October 4, 1965, the eighth anniversary of the Sputnik launch. The launch date was not specially chosen. It resulted from the combination of the celestial-mechanics possibilities and the requirement to coordinate this mission with the flights of the Zenit spy satellites and the next two Venera probes.

The launch was successful, as was the transfer in the direction of the moon. As always, there were problems with the I-100 unit. It had to be switched off frequently to prevent it from overheating again. This greatly complicated control and monitoring, especially in the last phase of the flight, but two hours before the planned landing, things looked good.

Finally the astronavigation system, which was supposed to position the probe for the landing burn maneuver, was activated. It was the system's task to locate the sun, the moon,

and the Earth. One of the optics had to have the moon in the center of its field of view; another, the Earth; and the third, the sun. Only if this happened could the probe be positioned so that radio altimeter's antenna showed exactly vertical to the surface of the moon, and determining the rapidly diminishing distance could begin. Then, at a precisely predetermined height, the system had to ignite the landing engine. If for some reason the Earth left the Earth sensor's field of view, the ignition would not take place.

There was great excitement in the control center in Simferopol. First came the report that the probe's vertical alignment had succeeded. This took place beginning at a distance of 3,000 miles above the moon's surface. At 2,400 miles, however, the Earth sensor lost its reference point, and the burn maneuver, which should have begun at an altitude of 46.5 miles, did not happen. All those present in the control center were able to perceive the impact on the moon. It was the moment in which transmission of the carrier wave ended.

The TASS communiqué of October 9 reported that "the automatic station Luna 7 has reached the moon in the area of the Sea of Storms." It continued: "While approaching the moon, the majority of the operations required for execution of a soft landing were carried out successfully. Several other operations were not carried out in accordance with the program and require further optimization. During the flight of Luna 7, a great amount of material was obtained for future operations."

On October 25, 1965, the Central Committee of the Communist Party and the Council of Ministers of the USSR issued a decree ratifying an additional circumlunar program completely independent of the manned moon landing. It had originally been envisaged that flying around the moon on highly elliptical Earth orbits, or the L1 program, should be an integral component of the L3 program. Now the L3 plan was encountering delays, and the prospects of carrying out this publicity-pregnant undertaking before the Americans were dwindling.

It was initially planned that Vladimir Chelomei's OKB-52 would carry out the new and independent L1 program completely on its own. That is what the first decree covering the L1 plan of August 3, 1964, had said. It was already developing its own manned transport system with the designation LK-1, which resembled the American Gemini capsules. It would be carried aloft by the Proton K booster, which was also envisaged for the lunar probes of the E-8 program. But now, on December 25, 1965, it was ordered that from OKB-52's Proton K/LK-1 project a combined project would be carried out by both organizations, using the OKB-1's 7K-L1 spaceship instead of OKB-52's LK-1.

But no matter who built which spaceships, the result was the same. In decrees issued in August 1964 and December 1965, the Soviet

Mission Data, Luna 7	
Mission designation	Luna 7
Date	October 4, 1965, 08:55 CET
Spacecraft	E-6 no. 11
Booster rocket	8K78 (U-103-27) Molniya
Spacecraft weight	3,315 pounds
Planned mission objective	Soft landing
Mission results	Failed shortly before the objective. Loss of position orientation immediately prior to ignition of the landing engine prevented its activation
Fate	Impacted at 9.8 degrees north and 47.8 degrees west (near Kepler crater) on October 8

Union, which in fact had barely enough resources for a single manned moon project, ordered two independent manned lunar projects that were not compatible with one another.

Contrary to the previous planning, Korolev's OKB-1 was now involved in the circumlunar L1 program. OKB-1 was to develop a variant of the Soyuz with the designation 7K-L1. It was a simplified version of the orbital Soyuz 7K-OK, which had no habitation module and was equipped with a simplified equipment unit with considerably less fuel.

Vladimir Chelomei's OKB-52 now had to develop only the booster rocket for the L1 program. It was the UR-500, which since its maiden flight on July 16, 1965, had borne the designation Proton K. The choice of the Proton K was also the reason for development of the Soyuz 7K-L1, since this rocket was not powerful enough to deliver a fully equipped LOK spacecraft of the L3 program, with its weight of almost 11 tons, to the moon. The weight of the L1 spacecraft was not supposed to exceed 6 tons. The Proton was capable only of placing that weight plus the 16-ton Block D into a low Earth orbit.

To reduce weight to that amount, the OKB-1 designers had to go to extremes. They removed the habitation module, which meant that the two cosmonauts who were to make the trip would have to spend seven days in the confined space of the landing module. They did away with the backup landing parachute and reduced the fuel supply of the service module, since returning from the translunar Earth orbit did not require a braking maneuver.

OKB-1 would also be responsible for development of the fourth stage of the L1 combination, the just-mentioned Block D. Its function was the same as Block L of the 8K78 in the Luna program at that time: it served as the transfer stage. Its task was to first give the probe spacecraft the decisive push to achieve a parking orbit around Earth, and then to place the L1 spacecraft on the highly elliptical Earth orbit for the circumlunar mission. This fourth stage was based on the N1's Block D and was one of the few common program elements used in both the L1 and the L3 programs.

With OKB-1 now involved in the circumlunar program, it had to develop three versions of the new Soyuz spacecraft independently of one another. There was the orbital version, which bore the designation 7K-OK (for Орбитальный Корабль, or "orbital ship" in English), and then there was the just-

The 7K-L1S was a special version of the 7K-L1 Zond. Just two examples of this special variant were built. They were used in the first and second test flights of the N1 rocket.

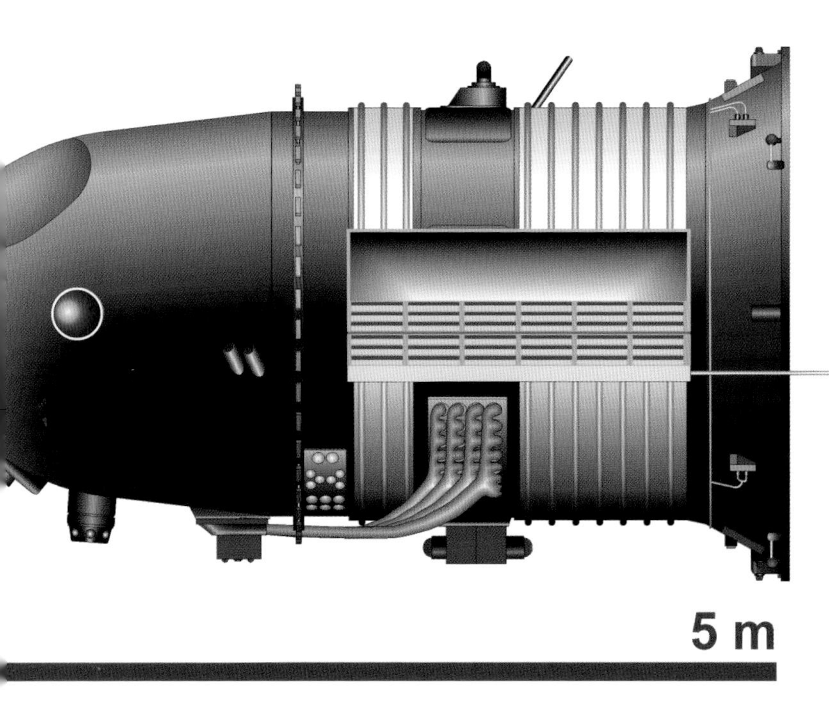

mentioned second version for the circumlunar flight with the designation 7K-L1, which had just a simplified service module and was supposed to fly without an orbital unit, and the 7K-LOK version for the moon-landing mission. These versions were so different from one another that they were in fact three completely independent manned spacecraft designs. Their only common feature was the reentry module, but even these exhibited considerable design differences. Given the already severe lack of resources, especially by OKB-1, this was a catastrophic decision.

Program delays soon arose in development of the Soyuz 7K-LOK, which were not solely due to technical problems. The political leadership of the USSR displayed uncertainty as to how the manned lunar program should be carried out, but it also revealed the technological gap that existed between the USSR and the US. There were enormous problems, especially in regard to development of the Igla rendezvous and docking system. This was because the Soviet engineers insisted on fully automatic systems and were not prepared to involve the cosmonauts in the

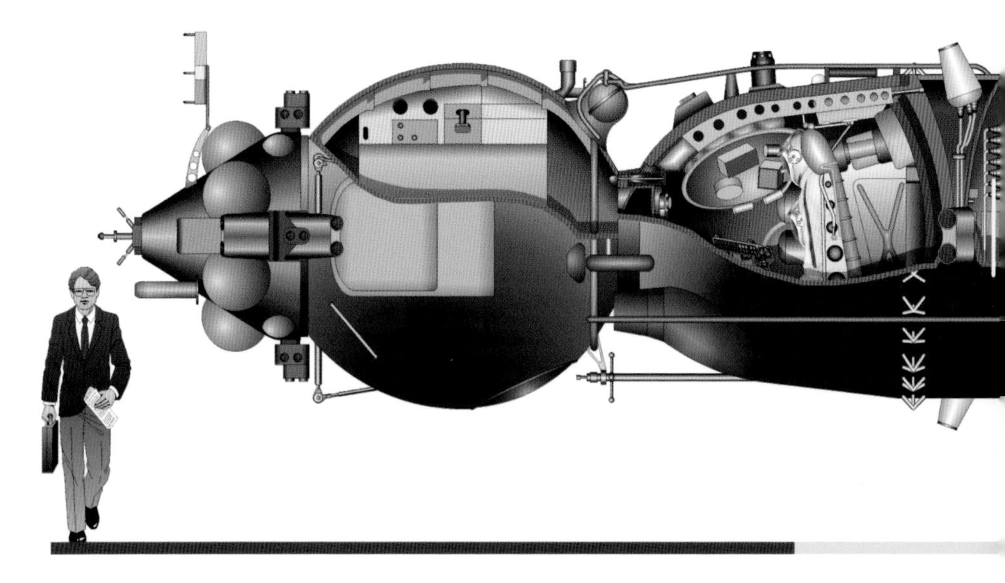

The 7K-L3 (also designated 7K-LOK) was envisaged for use in the manned moon-landing missions.

control and monitoring processes to the degree that the Americans were doing.

The first-generation Igla system could be used only in conjunction with two differently equipped Soyuz spacecraft. There was an active variant with the designation Igla 1 and a passive version called Igla 2. A Soyuz could accommodate only one or the other system. From the outset, therefore, one was defined as an active ship and one was passive. The active system alone required three different antenna arrangements: search antennas that could detect the passive partner at distances of 15.5 to 19 miles, a gyro-stabilized tracking antenna, which contacted the passive ship's transponder, and other special antennas for the various phases of the final approach. Each Soyuz spacecraft had to extend five antenna systems just for the rendezvous and docking process. Other antennas had to be deployed for all other purposes, such as communications and telemetry transmission. All in all, the early Soviet spacecraft had no fewer than twenty antennas onboard, and on one occasion an unnerved Korolev observed that instead of a manned spacecraft, he had designed an antenna carrier.

October 1965 was the month when Boris Chertok, head of the E-6 program, had to answer to the Council of Ministers for the continuing lack of success. With the help of Korolev, who made it clear to the ministers present that after a lengthy learning process they were just short of their objective, he was once again able to generate enthusiasm for the project.

On November 28, after three probes had been launched to Venus at brief intervals (only two of the three launches were successful), the launch of E-6 unit no. 12 and the eleventh

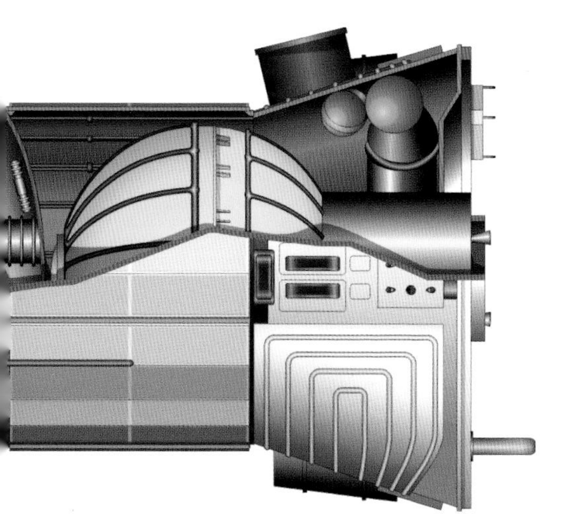

10 m

attempt to make a soft landing on Earth's satellite was scheduled for December 3, 1965. The launch was to take place at 13:46 Moscow time. The ballisticians had calculated that the probe would land on the moon at 23:57 on December 6.

It was to be Sergey Korolev's last mission. The launch, including the ignition of Block L, was faultless. TASS announced the new mission in the media and gave the probe the official designation Luna 8. The first contact with the probe took place as planned, as did the first corrective maneuver. There were minor irregularities in the course of later contacts. The actual burn periods deviated slightly from the estimated values, but this fault had no significant impact on the mission and was later corrected.

At 23:51 on December 6, the probe reached the exact point at which the preceding flight of Luna 7 had failed. But once again the crowning glory was not to be achieved. Pressurization of the airbags began thirteen seconds after the corresponding command was sent. The valve of the pressure tank for filling the airbags opened, and immediately the altimeter began losing orientation. Almost simultaneously the sensors also lost the sun and the Earth as reference points. The spacecraft had begun to tumble. The specialists calculated that the rate of rotation at the moment the retrorocket ignited was about 12 degrees per second. This was three times the value that the control system could compensate for.

The first impression that the flight controllers had was later confirmed during evaluation of the telemetry. A hole must have developed in one of the airbags, from which pressurized gas escaped, and must have changed the spacecraft's position in space so forcefully that the control system could not keep up. This was confirmed after closer examination. In inflated condition the airbags were held by a fiberglass structure that could easily break, forming sharp edges. One of these sharp edges had punctured the airbag, allowing the pressurized gas to escape. This had been overlooked during production and ground tests, in a clear case of negligence.

The summons by General Ustinov for a report on the latest failure was not long in coming. The heads of OKB-1 had begun preparing for it when fate struck, and this time in a very different form than the previous failed launch. As a result of this fate, the renewed defense of the program before the Council of Ministers never took place.

It began when on December 14, quite unexpectedly, Leonid Voskresenski died of heart failure at the age of fifty-one. Voskresenski had been one of Sergey Korolev's deputies for many years, until after 1963,

Mission Data, Luna 8	
Mission designation	Luna 8
Date	December 3, 1965, 11:46 CET
Spacecraft	E-6 no. 12
Booster rocket	8K78 (U-103-28) Molniya
Spacecraft weight	3,417 pounds (230 pounds for the landing unit)
Planned mission objective	Soft landing
Mission results	Failed immediately prior to objective. One of the airbags was damaged during the landing procedure. The escaping gas caused the probe to spin
Fate	Hard landing at 9.1 degrees north and 63.3 degrees west (in the Mare Procellarum) on December 6

following his first heart attack, he left the organization and headed a testing department in the Moscow Institute for Aviation Research, though he continued to function as an advisor to OKB-1. Voskresenski had played a major part in designing the N1, and he was a staunch advocate of a test stand for the entire first stage of the Russian moon rocket. He was, however, unable to impose his beliefs on Korolev.

He knew that the construction of such a test stand not only would be very expensive but would also mean the program standing still for years. The future would show that Voskresenski was ultimately right. Korolev was one of the speakers at Voskresenski's funeral. Immediately after the burial, Korolev was admitted to the hospital for examination on account of rectal bleeding. There, an operation was scheduled for mid-January to remove what the doctors had diagnosed as benign bowel polyps. They said that Korolev would have to give up his responsibilities for only about a week.

Many decisions concerning the L3 program were made in the last days of 1965. Most were of a technical nature, but there were also a number of programmatic decisions. The flight of a moon landing using an N1 booster rocket was planned in detail. The central element with which this was to be achieved was the LK lander. Korolev's calculations revealed that this element, even with an N1 payload capacity of 105 tons, could not weight more than about 6 tons. By comparison, NASA's lunar lander weighed between 16.5 and 18.75 tons. If one took into account the generally poorer construction of the Soviets, the heavier electronic components, and the much less capable computers in the Soviet Union, it was a nut that was almost impossible to crack. In any case the decision to put just one cosmonaut on the moon made it somewhat easier, even though this decision itself raised new problems, especially with regard to the cosmonaut's safety. Discussions went on for a long time about what would happen if the lone space traveler crashed on the moon and no one could come to his aid.

Unlike the American two-stage lander, the Soviet solution was in principle a single-stage vehicle. Only the landing gear remained on the moon (unlike the American solution, in which the entire descent stage was left on the surface). The lone cosmonaut took off with the same engine with which he had landed.

The procedure was very risky. The Americans had also examined it and rejected it as too dangerous. The landing procedure envisaged almost the entire descent, to about 6,500 feet, being carried out by Block D, the same propulsion unit that had slowed the LOK orbital unit and the LK lander into lunar orbit. Afterward it would be separated and crash onto the surface of the moon. The remaining descent and the landing, and the subsequent

ascent, could then be carried out with a relatively small engine. Here, however, the Soviets allowed themselves the luxury of carrying a reserve engine with the designation RK-859.

Both the primary and reserve engines with the designation RK-858 were very modern systems that were capable of varying their thrust by about 60 percent. After jettisoning of Block G at an altitude of about 1.2 miles, the remaining speed (about 300 feet per second) and altitude had to be degraded in fifty to sixty seconds. Afterward the cosmonaut was left with a maximum of another ninety seconds (by later calculations, just fifteen to twenty seconds) to find a place to set down and land. Then for his ascent, he had available a burn duration of 350 seconds. That would have been sufficient to carry out an active docking with the LOK spacecraft in orbit.

The two combustion chambers of the reserve engine were located to the right and left of the primary engine. Both systems were activated for launch. The reserve engine continued running until the onboard computer had completed a diagnosis of the primary engine. If the result was positive, then the reserve engine was shut down again and only the primary engine continued running. The reserve engine would, however, be capable of reigniting within three seconds should the primary engine develop a problem.

The third component of the L3 combination, after Block G and Block D with the LK unit, was the lunar orbital vehicle, or LOK. It was a drastically modified version of the 7K-OK version of the Soyuz spacecraft. The standard Soyuz had three main components; namely, the spherical habitation module, the landing module in which the cosmonauts would return to Earth, and the instrumentation and propulsion section. The LOK spacecraft also included the Block I engine, with which it would leave lunar orbit on the Earth transfer trajectory.

The habitation module had two hatches. One hatch led to the return module, and the second hatch, on the side, was an exit door, through which a cosmonaut could move into open space. The passageway to the return module was hermetically sealed. The unit, with a diameter of 7.4 feet, assumed the function of an airlock. The cosmonaut who would land on the moon had to carry out an extravehicular activity to reach the LK lander. The OKB-1 designers had also dispensed with a pressurized hatch connection between the two spacecraft. Such a pressurized tunnel would have resulted in several hundred pounds of additional weight, which they simply didn't have.

The Block I engine was the main deviation from the original Soyuz design. It was in a skirt-shaped enlargement at the rear of the ship. The rocket engine had a single purpose: to take the two cosmonauts out of lunar orbit and place them on course for home. This two-chamber engine produced 76,435 pounds of thrust. In the rear of the LOK spacecraft there was also a single standard Soyuz main engine with 9,440 pounds of thrust, which could be ignited up to thirty-five times. This engine was to be used for all the course corrections needed on the way back to Earth.

Both engines drew their fuel from the two massive—almost 6 feet long—two-part cylindrical fuel tanks inside the instrumentation section. These engine units were supported by sixteen attitude control thrusters, which also took their fuel from the large central tank. The entire LOK was considerably larger than the standard Soyuz, fairly close to 33 feet. It did have almost the same diameter as the Soyuz, 9.6 feet, and it weighed 10.85 tons in lunar orbit.

The fourth element of the N1-L3 combination was the rescue tower. It was very similar

LK lunar lander. Views from two sides.

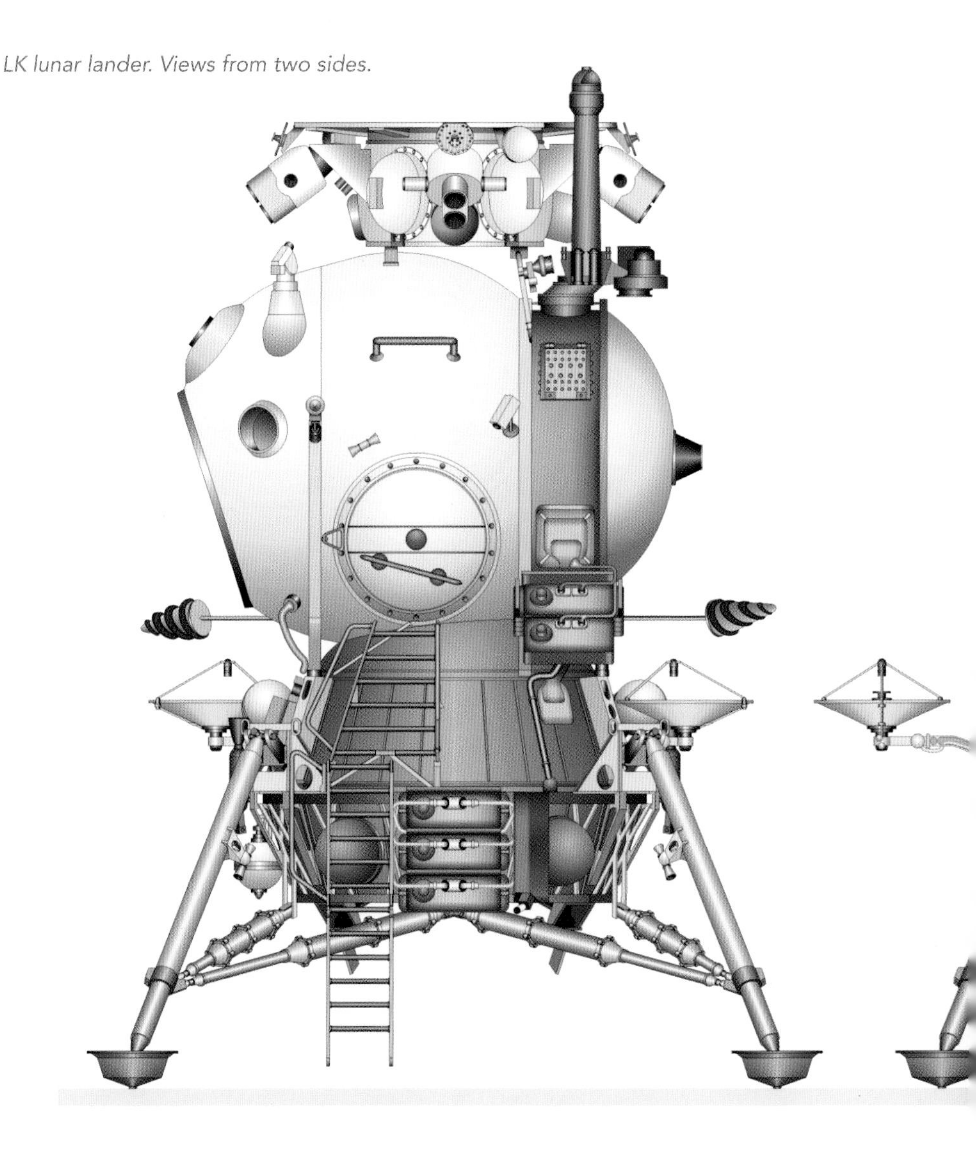

in design to that of the Soyuz 7K-OK and, like it, consisted of two solid-fuel rockets, the same rescue rockets that initially separated the payload fairing with the orbital module and the return cabin, but not the instrumentation section, from the rocket. The second propulsion system was the so-called separation engine, which pulled the fairing away from both these components. The N1's rescue system was clearly more capable than the corresponding system of the orbital Soyuz. For one thing, it had to be capable of dealing with the increased mass of the LOK habitation module and the return cabin, and it also had to be capable of transporting both these components significantly farther and higher, since in the event of an explosion of an N1, the danger area would be much greater than that of a Soyuz booster rocket.

What would a nominal N1-L3 mission have looked like? Ideas about this changed

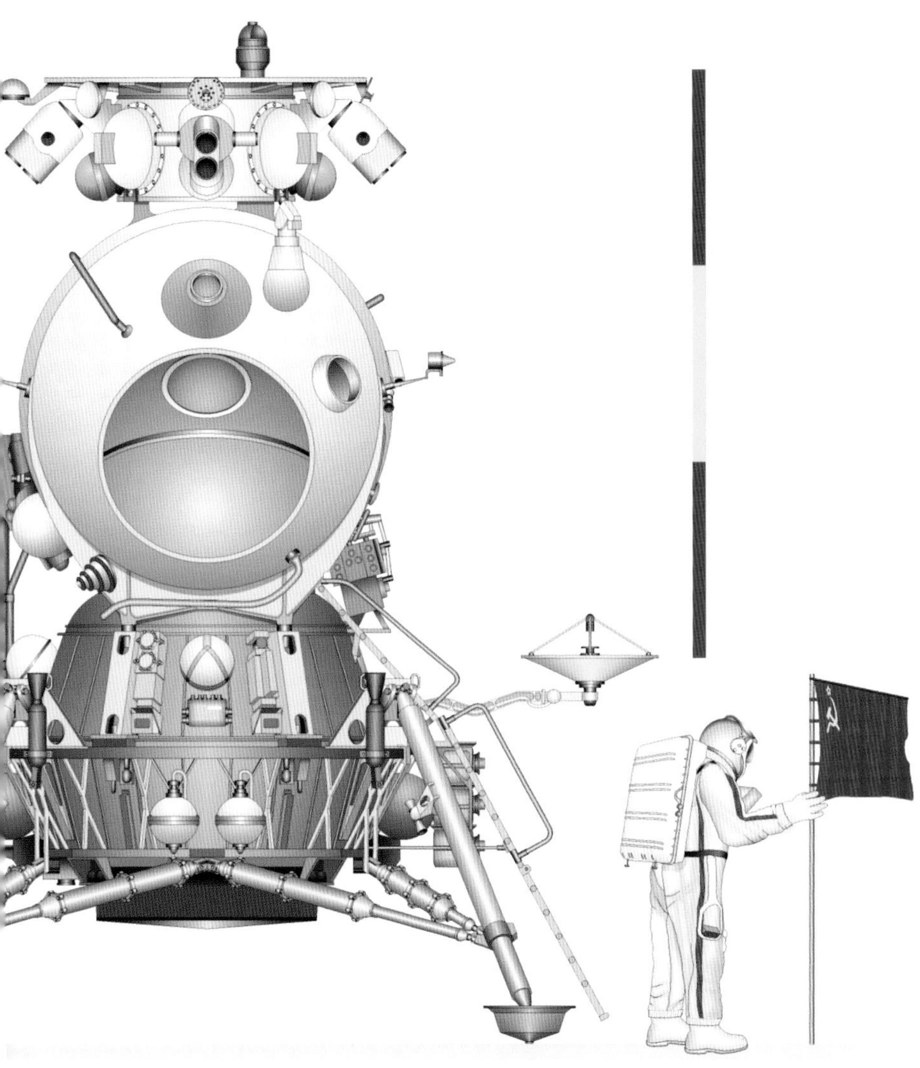

continually in the years 1965–69, but the principal elements of the plan were roughly as follows.

The 3,110-ton booster rocket with its approximately 105-ton payload would launch with its crew of two from Launchpad 110 at Tyuratam. During the Block B burn phase the huge fairing and the launch rescue system would be jettisoned. Nine minutes after leaving the launchpad, the Block V stage would place the entire L3 unit into an almost circular Earth orbit, with an orbital inclination of 51.6 degrees to the equator, at an altitude of 137 miles.

This would be followed by a twenty-four-hour flight phase, during which the entire system was put to the acid test. During the seventeenth orbit of Earth, the engine of Block D was to fire for eight minutes at the predetermined point, placing the L3 combination into an extremely high elliptical orbit that

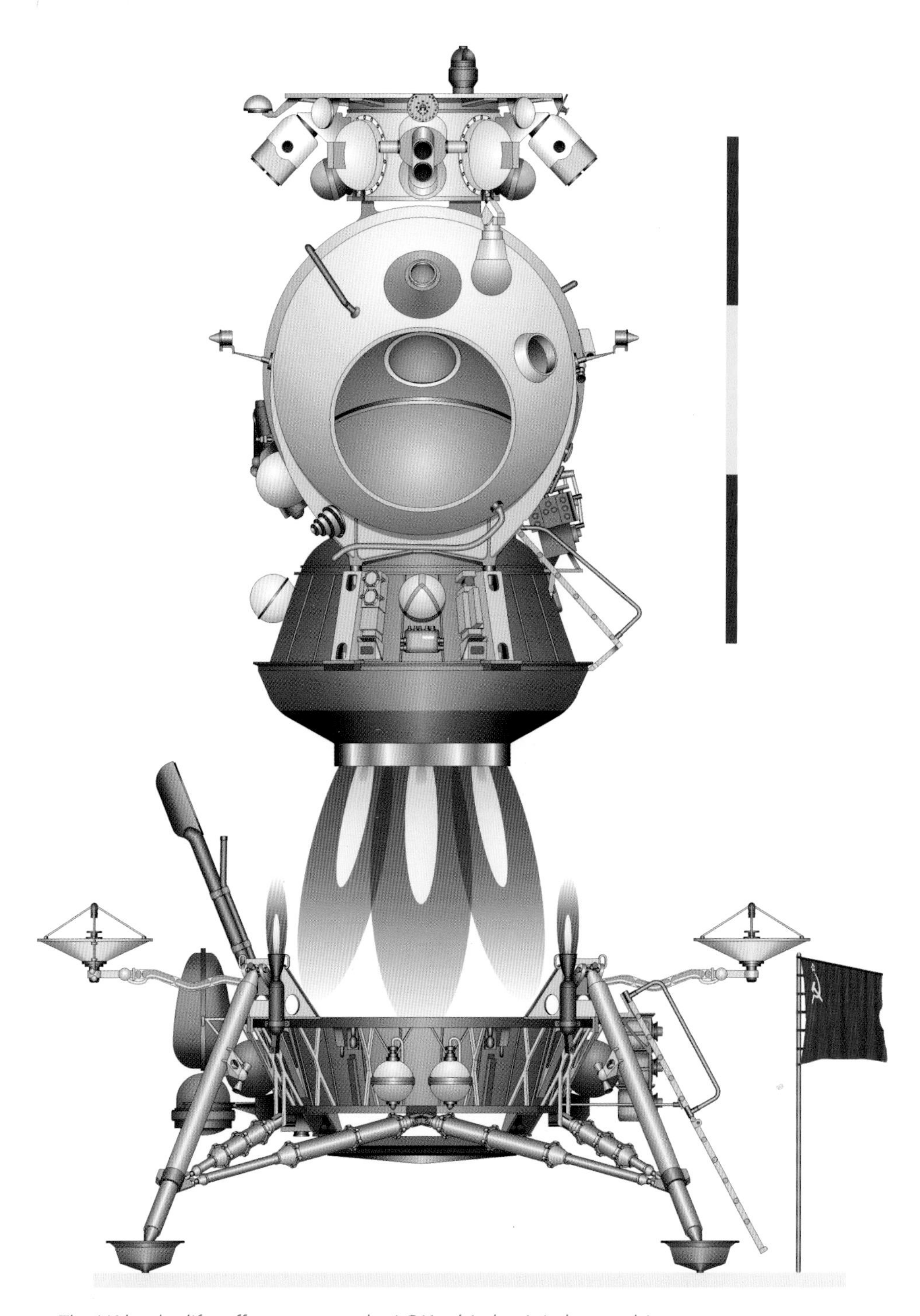

The LK lander lifts off to return to the LOK orbital unit in lunar orbit.

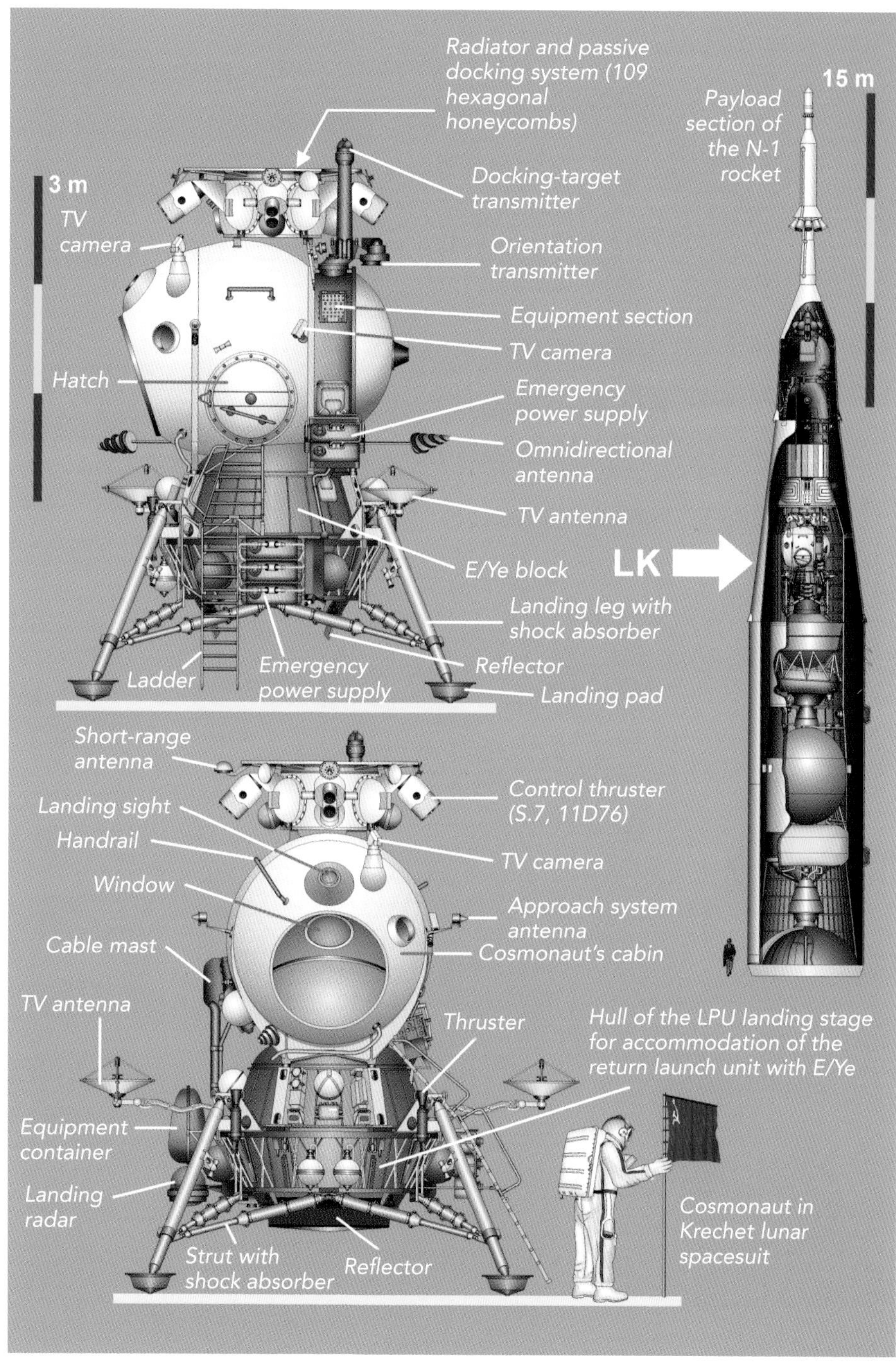

LK lunar lander (Product 11F94).

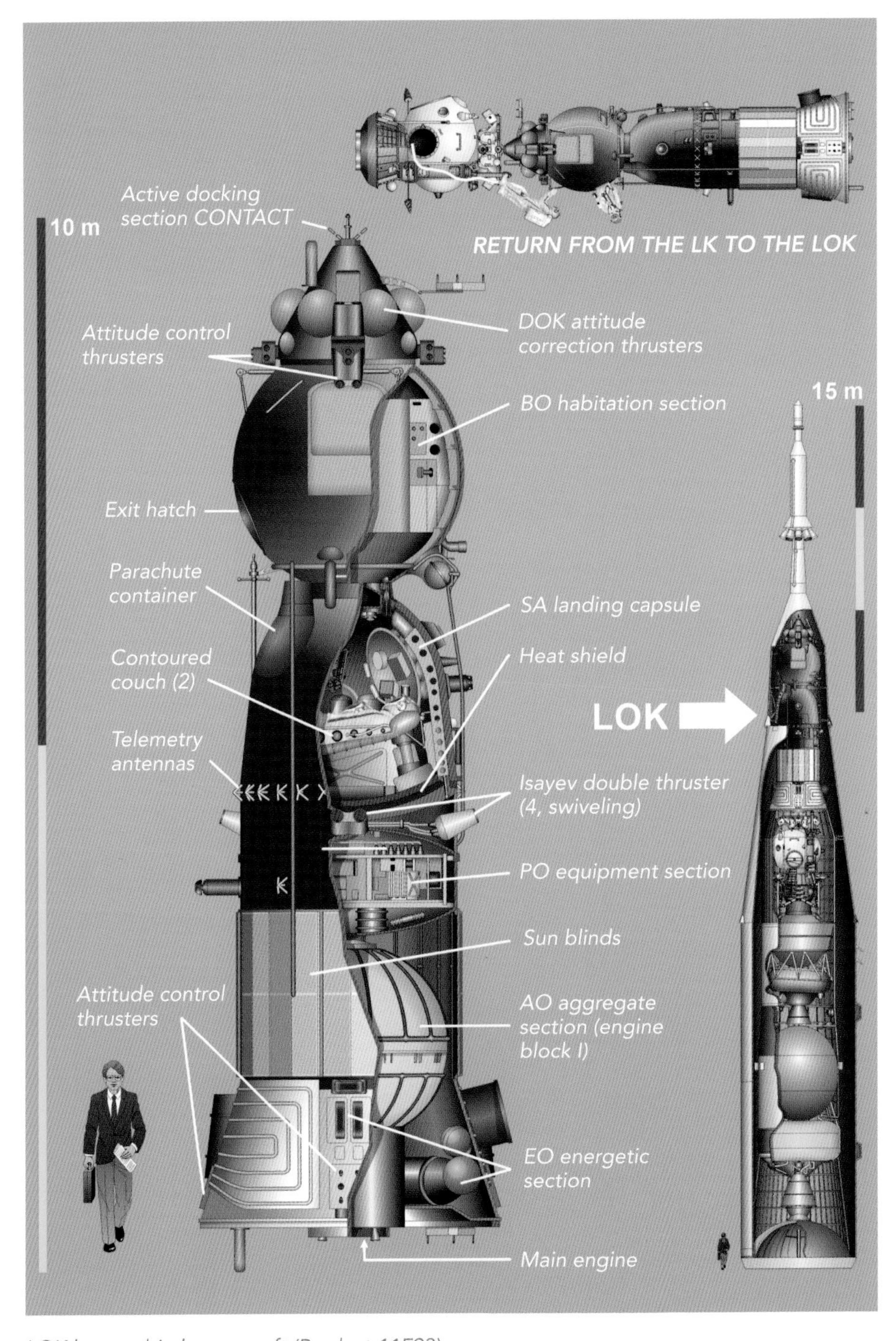

LOK lunar orbital spacecraft (Product 11F93).

enabled a translunar flight on a free return trajectory. A few minutes after burnout, Block G would be jettisoned.

Should some easily rectifiable problem appear during the seventeenth orbit, then the translunar burn maneuver could be tried again two orbits later.

The transfer to the moon would take 101 hours. Two course corrections were scheduled during this time, both carried out by Block D. The first course correction was supposed to take place eight to ten hours after insertion into the transfer trajectory, with the second occurring twenty-four hours before veering into lunar orbit. During the first course correction the L3 combination also left the free return trajectory to Earth.

After a transfer phase lasting the better part of four days, the combination approached the moon and, assisted by gravity, swung around Earth's natural satellite. On the reverse side of the moon, Block D fired for several minutes and placed itself, the lander, and the LOK orbital ship into lunar orbit with an apogee of about 125 miles and a perigee of 90 miles. Block D ignited two more times during the fourth and fourteenth orbits, lowering the orbit to an apogee of 60 miles and a perigee of just 12 miles.

The crew would then check the LOK spacecraft's orbital module and all systems of the combination of the LOK, LK lander, and Block D. At this point in the mission the LK lander would still be inside the cylindrical adapter section, a sort of second payload fairing, which provided the combination's stability and was part of the L3 complex.

The commander, protected by his Krechet spacesuit, would now leave the habitation module. With the help of a mechanical arm, he would be hefted to the hatch on the inner fairing, beneath which was the lander. This aspect is both original and curious, since obviously the Soviets did not trust their cosmonauts to cover the distance of about 60 feet from the outer shell of the L3 combination to the hatch themselves. One must, however, bear in mind that apart from Leonov's adventurous twelve minutes in space during the Voshkod 2 mission in 1965, prior to 1969 the Soviets had no experience in extravehicular activities.

Upon reaching the hatch on the fairing, he would first open it, then a second leading into the LK lander. The flight engineer would still be in the LOK habitation section. Wearing an Orlan spacesuit, he would monitor the transfer of his commander and intervene if this proved necessary. The commander would meanwhile arrive in the LK lander, close the hatch, pressurize his spacesuit, and begin checking the systems.

The adapter panels would then be separated, thus exposing the lander to space for the first time. When the combination began to overfly the landing area, Block D would fire for the last time. The Planeta landing radar would then initiate an automatic descent, taking the lander to a height of about 6,500 to 4,900 feet above the surface of the moon. At that point the almost completely burned-out stage would be shut down and jettisoned, subsequently crashing near the landing site.

The lander's engine would then fire, controlled by the commander, who could manually adjust the thrust. About a minute after separation of Block D, he would still be a few dozen yards high in hovering flight. He would then have about twenty-five to thirty seconds to look for a good place to land and then begin the final descent. In the best-case scenario, the descent to landing, from activation of the lander's engine until touchdown, would take about sixty to seventy seconds. After this, about thirty more seconds of reserve time would be available. If no place to land could be found by that time,

Mockup of an LK lander. On display at the Science Museum in London as part of a special display in 2016. Photo: Ute Gerhardt

Mockup of an LK lander. On display at the Science Museum in London as part of a special display in 2016. In the background is a Lunokhod rover. Photo: Martin Trolle Mikkelsen

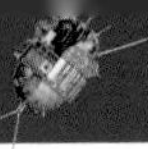

the landing would have to be aborted. The commander would then engage full thrust and fly back to low lunar orbit, where he would be controlled from the LOK.

The cosmonaut's first tasks after touching down would be to check his vehicle's systems, check his Krechet spacesuit, reduce pressure in the lander, and leave it by way of the small oval hatch in its side. Watched by a television camera, he would then climb down the ladder to the ground. On the surface he would plant a Soviet flag, unload several scientific experiments, collect soil samples, and take photos. After spending about ninety minutes on the surface, he would then climb back into his spacecraft.

At the predetermined time, all connections to the landing gear would be separated pyrotechnically, and the vehicle would head back into lunar orbit. There the cosmonaut would wait in the lander, since in the following rendezvous maneuver the LOK orbiter would have the active role. In the nominal case, docking would be fully automatic and take place with no intervention by the crew with the aid of the contact radar. There was, however, a manual backup solution, in which the flight engineer on the LOK manually carried out the rendezvous and docking. Maneuvering in orbit would be carried out with the attitude control thrusters and the Soyuz standard engine. After docking, the commander would make another extravehicular activity, returning to the LOK and taking with him the soil samples he had collected.

During the thirty-eighth orbit of the moon, the lander and the orbital module would separate. Today there is some debate about this point, and it is depicted differently in the literature and period drawings. Some sources claim that the orbital module would remain on the LOK and be placed onto the Earth transfer trajectory with it. Because of the uncommonly tight weight of the L3 system, this seems rather improbable, since at least 1,100 pounds of fuel would be required to accelerate the roughly 1.1-ton orbital unit out of lunar orbit to the required velocity of 3,600 feet per second (almost 2,400 mph).

During the thirty-ninth orbit the Block I engine would be fired on the reverse side of the moon and place the LOK back on course for Earth. Total time in lunar orbit was limited to seventy-seven hours. The return flight to Earth lasted eighty-two hours. Two course correction maneuvers would be carried out with the Soyuz engine, one about twenty-four hours after moon escape maneuver, and the second about forty-four hours later.

Approaching the Earth, about two hours prior to reentry, the propulsion and equipment

One of the LK lander's landing pads.

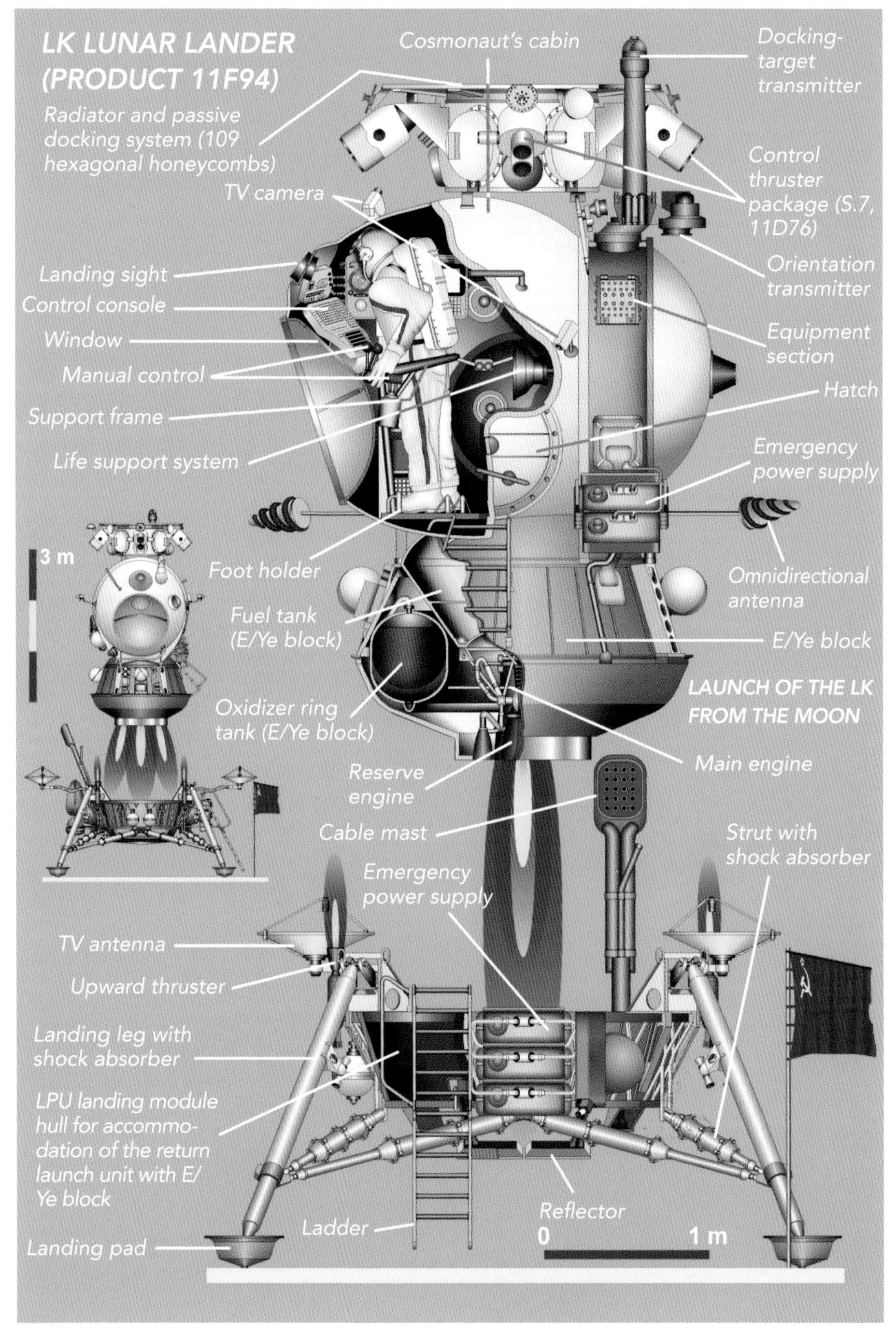
LK LUNAR LANDER (PRODUCT 11F94)
Radiator and passive docking system (109 hexagonal honeycombs)
Cosmonaut's cabin
Docking-target transmitter
TV camera
Control thruster package (S.7, 11D76)
Orientation transmitter
Landing sight
Control console
Window
Manual control
Equipment section
Hatch
Support frame
Life support system
Emergency power supply
3 m
Foot holder
Omnidirectional antenna
Fuel tank (E/Ye block)
E/Ye block
LAUNCH OF THE LK FROM THE MOON
Oxidizer ring tank (E/Ye block)
Main engine
Reserve engine
Cable mast
Strut with shock absorber
Emergency power supply
TV antenna
Upward thruster
Landing leg with shock absorber
LPU landing module hull for accommodation of the return launch unit with E/Ye block
Reflector
Ladder
0
1 m
Landing pad

module would be separated. Reentry itself would be made at a very shallow angle to reduce forces on the crew and to minimize heat buildup on the heat shield. The Americans, who also used this maneuver in the Apollo program, called it the "double skip." The spacecraft enters Earth's atmosphere, experiences the first burn phase, and is controlled aerodynamically in such a way that it briefly leaves the atmosphere before ultimately reentering it. The landing by parachute would then take place in Kazakhstan.

1966: FIRST HALF OF THE YEAR

On January 14, 1966, Sergey Korolev died during an operation in a Moscow hospital. On the one hand, for the first days after his death, and around the obsequies, there was total paralysis, while on the other hand the plans for the next activities were so far advanced that they could initially continue even without Korolev's aid, as if nothing had happened. It was not until the weeks and months after his death that it became more obvious that the driving force behind the Soviet lunar program was gone.

In December, January 31, 1966, had been set as the date for the twelfth attempt to make a soft landing on the moon. The launch at 12:42 Central European Time that day was successful, and soon the probe was on its way to the moon, and the TASS news agency named it Luna 9. This time everything worked. The spacecraft stabilized itself as planned at an altitude of 5,157 miles. Vertical positioning, which had been the point of failure for the Luna 8 probe, was completed without error.

At an altitude of exactly 248,966 feet above the surface of the moon, the two side modules, weighing 686 pounds, which were not required for the landing, were jettisoned. This included the landing antenna, which had done its job. The Isayev S5.5A main engine, with its thrust of 5 tons, ignited as planned and brought Luna 9 down precisely. When the electronics had registered the planned cumulative change in velocity, the engine shut down. This happened at a height of about 750 feet. This left only the four stabilizing engines attached to the outriggers. At a height of 15 feet a sensor reported ground contact and also switched these engines off. That was the signal for the probe to jettison the landing capsule, which was 23 inches in diameter. It fell toward the ground at a velocity of 13 miles per hour and after bouncing several times came to a stop and opened its four "wings."

The landing took place at 19:45:04 Central European Time. Then, exactly 258 seconds after the capsule had come to rest, a timing device activated the radio transmitter. Nineteen days after Korolev's death, the Soviet Union had achieved a new first on the moon.

Seven hours after the landing, Luna 9 transmitted the first pictures to Earth. Because of the usual secretiveness of the Soviet leadership, it was several more hours before the internal release of the photographs. This was enough time for the Jodrell Bank Observatory to score a special public-relations coup. As in years past they had observed this Soviet moon mission closely and had intercepted Soviet radio signals. After the landing on the moon, one of the engineers realized that some of the signals were coming in using the internationally recognized Radiofax system. In the 1960s

The landing area for Luna 8 and Luna 9.

this was a system used by the press to transmit photos by radio. A call to London's *Daily Express* was sufficient, and one of the journalists rushed to Jodrell Bank with a reproducer. There the signal was decoded and spread worldwide, even before the flabbergasted and angry Soviets could do it themselves.

Now there remained another lunar first to be achieved: the world's first lunar-orbiting probe. The Soviet engineers and technicians modified a Luna 9 probe into the E-6S series. The landing capsule was replaced with a small orbital unit without position control equipment, weighing about 540 pounds. In this case the main engine was not used for landing, but instead to slow the probe into an orbit around the moon. The first attempt to send such a modified E-6 unit to the moon failed on March 1, 1966, once again because of the Block L stage of the 8K78M booster rocket. The attitude control system failed and the combination of upper stage and space probe remained in Earth orbit. It was given the innocuous designation Cosmos 111.

Mission Data, Luna 9	
Mission designation	Luna 9
Date	January 31, 1966, 12:45 CET
Spacecraft	E-6 no. 13
Booster rocket	8K78M (U-103-32) Molniya M
Spacecraft weight	3,483 pounds (218 pounds landing weight)
Planned mission objective	Soft landing
Mission results	First successful soft landing on the moon in the history of spaceflight at 19:45 CET on February 3. Transmitted nine images of the lunar surface
Last contact	February 6, 1966, 23:55 CET
Fate	Landed at 7.08 degrees north and 64.37 degrees west (near Reiner and Marius craters)

The space probe Luna 10 (here, a model) and Cosmos 111 were identical in construction.

March 16, 1966: Gemini 8 with astronauts Neil Armstrong and David Scott carries out the first docking maneuver in space. Immediately afterward, there is a short circuit onboard, and Gemini 8 becomes the first American spacecraft that must break off a mission prematurely. It lands safely in the Pacific.

In March 1966, there was a major restructuring of industrial complexes in the Soviet Union with associations to the military. OKB-1 became the "Central Design Bureau for Experimental Machine Building," abbreviated TsKBEM. By that time the manned lunar program was already in very bad shape. No production drawings existed for any of the key elements, not for the LOK lunar orbiter, for the LK Lander, or for elements of the N1 rocket. It had been hoped that the accelerated circumlunar program would yield better results; however, its technical concept was far simpler, and much smaller in magnitude in terms of the necessary resources. But it also appeared that no hardware for the 7K-L1 would be ready to fly before the end of

Mission Data, Cosmos 111	
Mission designation	Cosmos 111
Date	March 1, 1966, 12:04 CET
Spacecraft	E-6S no. 204
Booster rocket	8K78M (U103-41) Molniya M
Spacecraft weight	3,492 pounds
Planned mission objective	Lunar orbiter
Mission results	Achieved low Earth orbit only. Ignition of Block L upper stage failed
Fate	Achieved an orbit of 118 x 140 miles and 51.85 degrees inclination. Crashed back to Earth on March 3

1966. The only component already on the test stand at that time was the 7K-OK, the Earth orbit version of the Soyuz.

Thirteen days were planned for testing of the first unit of the new spacecraft. Then it was supposed to go to Baikonur, where it would undergo preparations for the first unmanned test mission. In fact the tests lasted 112 days, or 2,240 hours. The checks revealed 2,123 defects and that 897 modifications were necessary. Only after these had been carried out could Flight Unit 1 be delivered to Baikonur. Another 300 problem areas were discovered there in the two months of preparation for the first flight. Ninety percent of these defects were system errors in the new Soyuz design and not problems with the individual vehicle being tested. It was realized that Korolev had neglected to have engineering and qualification models built, now standard practice for new spacecraft. Such a systematized process would have extended the schedule by a year, however, and one or two additional examples of the Soyuz would have had to be built for this purpose. Afterward, however, if further errors were found they would have been limited to the single production model of the spacecraft and not to the layout and basic design. However, resources were so limited and management's obligations to governmental requirements, which were in fact impossible to meet, were so strict that this commitment of time and resources was regarded as indefensible.

The mission for an unmanned first flight was very complexly laid out. A second launch was to follow just two days after the launch of the first Soyuz from Launchpad 31, and during the first or second orbit it was to dock with the first Soyuz. If this succeeded, a second pair of Soyuz spacecraft would repeat the maneuver with a crew onboard.

Testing of the 7K-OK no. 1 began in May, and one month later tests on the second flight unit began in parallel. Flight units 3 and 4 were checked out between September and November. The number of faults dropped over time but was still very high. Flight unit 4 had 736; unit 5 had 520. Testing times also became shorter—testing time for the fourth flight unit was only about a third of the time that had been required for the first unit.

On March 31, 1966, the backup unit for the lunar orbital probe stranded in low Earth orbit on March 7 was launched. This time everything went according to plan. After the translunar burn maneuver by the Block L stage, the vehicle was given the name Luna 10. The course correction maneuver was completed without problem, and on April 3 Luna 10 became the first spacecraft to enter orbit around the moon. Then the "bus" separated from the orbiter. The orbit had a perigee of 218 miles, an apogee of 621 miles, and an inclination to the lunar equator of 71.9 degrees.

Luna 10 had no photographic equipment onboard. There would have been no point to this, however, since it had no attitude control system. It did, however, gain some scientific fame, because Luna 10 found the first so-called mascon, a concentration of denser material beneath the lunar surface, which has a significant influence on the tracks of spacecraft orbiting the moon.

The probe carried with it a playback device that played a disc with the Internationale, the battle hymn of the international worker's movement, during the Twenty-Third Congress of the Communist Party of the Soviet Union (which took place between March 29 and April 8, 1966). The spacecraft, which was powered solely by chemical batteries, remained in operation until April 30.

Model of Luna 10's orbital unit.

Another view of Luna 10's orbital unit.

May 30: The Americans launched the moon-landing probe Surveyor 1. The vehicle was a complete success, landing on the Mare Procellarum on June 2. Contact was maintained with the probe until January 7, 1967. It transmitted its last photograph (of a total of 11,240) on July 13, 1966.

June 3: The US carried out the mission of Gemini 9 with astronauts Stafford and Cernan onboard. Several successful rendezvous maneuvers and one less successful extravehicular activity were carried out.

June 18: During the mission of Gemini 10 with John Young and Michael Collins, docking maneuvers and extravehicular activities were carried out. The docking was successful, but there were again problems during the EVA.

Mission Data, Luna 10	
Mission designation	Luna 10
Date	March 31, 1966, 11:48 CET
Spacecraft	E-6S no. 206
Booster rocket	8K78M (U-103-42) Molniya M
Spacecraft weight	3,488 pounds (orbiter: 540 pounds)
Planned mission objective	Lunar orbiter
Mission results	First lunar orbiter in the history of spaceflight Orbit 217 x 621 miles, inclination 71.9 degrees
Last contact	May 30, 1966
Fate	Crashed on the moon, date unknown

1966: SECOND HALF OF THE YEAR

Mission Data, Luna 11	
Mission designation	Luna 11
Date	August 24, 1966, 09:09 CET
Spacecraft	E-6LF no. 101
Booster rocket	8K78M (U-103-43) Molniya M
Spacecraft weight	3,615 pounds
Planned mission objective	Lunar orbiter
Mission results	Lunar orbiter. TV camera could not be used because of orientation problems with the probe
Last contact	October 1, 1966
Fate	Crashed on the moon, date unknown

August 10: The US launched the orbital probe Lunar Orbiter 1. The mission was highly successful, and 229 high-resolution photographs of potential Apollo landing sites were taken. On October 29, the probe was intentionally crashed onto the surface of the moon so as to cause no interference with the following lunar orbiter probes.

On August 24, 1966, the Soviet Union launched the moon probe designated Luna 11. This space probe was also based on the E-6 platform, but its design was more technically advanced than that of Luna 10. The launch and transfer to the moon were uneventful, as was the swing into lunar orbit at 22:49 Central European Time on August 27. The space probe entered an orbit with a perigee of 99 miles and an apogee of 742 miles. The transmission of pictures was unsuccessful, however, because a foreign object in the attitude control system prevented the camera from being pointed at the surface of the moon. The transmission of physical data proceeded as planned, however. This probe was also equipped solely with chemical batteries, whose operating life was very limited. The last data transmission from Luna 11 was received on October 1, 1966.

September 12: During the mission of Gemini 11 with astronauts Conrad and Gordon, several docking maneuvers were carried out and the spacecraft reached a record altitude of 683 miles. An extravehicular activity by Gordon was again less successful.

September 20–23: The mission of the landing probe Surveyor 2 was a failure because one of its engines failed during a course correction maneuver prior to landing. The probe crashed in the Sinus Medii.

By October the interruption in manned flights by the Soviets had reached one and a half years. In that time the Americans had launched no fewer than nine Gemini spacecraft, placing twenty-one astronauts into orbit. Three of them had even flown twice. The Soviet Union had put eleven cosmonauts into space. The total number of manned flights was now nine by the Soviet Union compared to fifteen by the Americans. And the next manned Soviet flight faced an uncertain future.

Georgi Tyulin, the deputy minister of the Ministry of Machine Building, into whose area

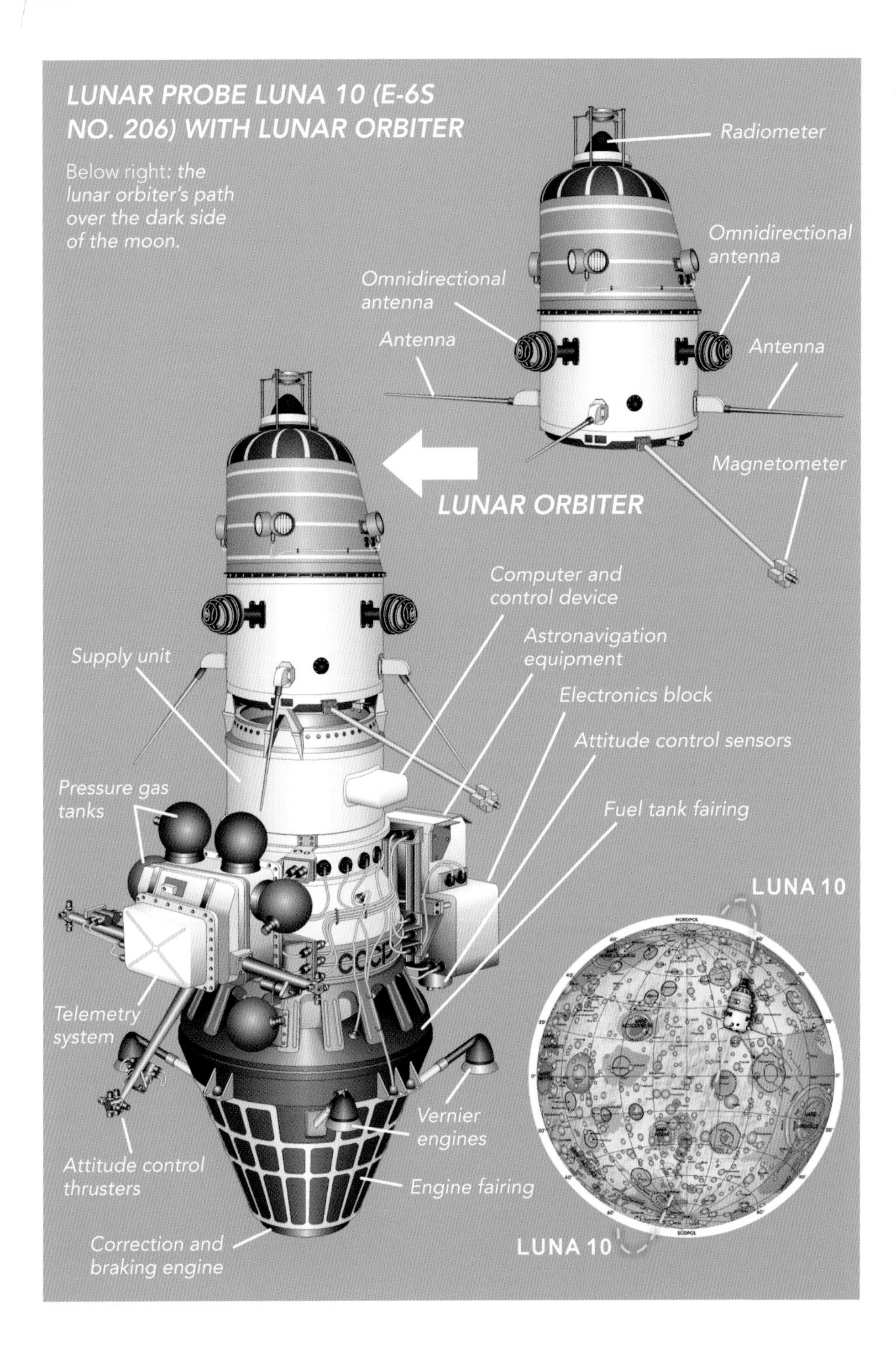
LUNAR PROBE LUNA 10 (E-6S NO. 206) WITH LUNAR ORBITER
Below right: the lunar orbiter's path over the dark side of the moon.
Radiometer
Omnidirectional antenna
Omnidirectional antenna
Antenna
Antenna
Magnetometer
LUNAR ORBITER
Computer and control device
Astronavigation equipment
Electronics block
Attitude control sensors
Fuel tank fairing
Supply unit
Pressure gas tanks
Telemetry system
Attitude control thrusters
Vernier engines
Engine fairing
Correction and braking engine
LUNA 10
LUNA 10

of responsibility the former OKB-1 and now TsKBEM fell, told Vassily Mishkin, Korolev's successor, that the government expected a manned flight around the moon as a "gift" to the celebrations of the fiftieth anniversary of the October Revolution in November 1967, and a landing on Earth's natural satellite the following year. Mishkin could not make this promise, however. Instead, he promised a manned flight by two Soyuz spacecraft by October, with a docking maneuver between the two units. Thus, he argued, they could at least trump the Americans in this area.

On October 22, 1966, the Soviet scientists sent another lunar orbiter, Luna 12, on its way to Earth's satellite. It entered orbit successfully on October 25. The initial orbital parameters showed a perigee of 62 miles, an apogee of 1,081 miles, and an inclination to the lunar equator of 10 degrees. This probe was also exclusively battery powered and therefore had a very limited operating life. It transmitted physical data and an unknown number of photographs until January 19, 1967. At the time, the Soviets published just a single photo showing the Sea of Rains and the crater Aristarchus. Resolution of this photo was 49 to 65 feet per pixel.

In the days leading up to the first unmanned Soyuz launch, there was a certain level of self-confidence among those responsible and the technicians. Discussions began regarding the composition of the two crews for the subsequent manned double mission. The first Soyuz was to launch with a single cosmonaut; the second, just one day later, with three crew members. The first spacecraft was to be the active ship, which had to carry out the rendezvous and docking maneuver. The second ship played the passive role.

Vassily Mishkin tried hard to push through a cosmonaut candidate from the OKB-1 cadre for the primary crew of the second Soyuz, either Alexey Yelisiyev or Viktor Garbatko. Only the primary pilot of the active Soyuz was finalized and not up for discussion: Vladimir Komarov, who had already commanded the Voshkod 1 flight. The crew's task was a crew change after docking. Two cosmonauts from the second Soyuz spacecraft would carry out an extravehicular activity and join the lone cosmonaut in the first ship and then return to Earth with him. This maneuver was a simulation

November 6, 1966: The US launched the orbital probe Lunar Orbiter 2. The mission was very successful, and 819 high-quality pictures were taken of potential landing sites. The probe conveyed data until it was intentionally crashed on October 11, 1967.

November 11–15, 1966: The mission of Gemini 12 with astronauts Lovell and Aldrin concluded the Gemini program. The flight was extraordinarily successful, and in addition to rendezvous and docking, for the first time it included a completely successful extravehicular activity.

Mission Data, Luna 12	
Mission designation	Luna 12
Date	October 22, 1966, 09:38 CET
Spacecraft	E-6LF no. 102
Booster rocket	8K78M (U-103-44) Molniya M
Spacecraft weight	3,571 pounds
Planned mission objective	Lunar orbiter
Mission results	Lunar orbiter
Orbit	82 x 745 miles; sent back an unknown number of photographs
Last contact	January 19, 1967
Fate	Crashed on the moon, date unknown

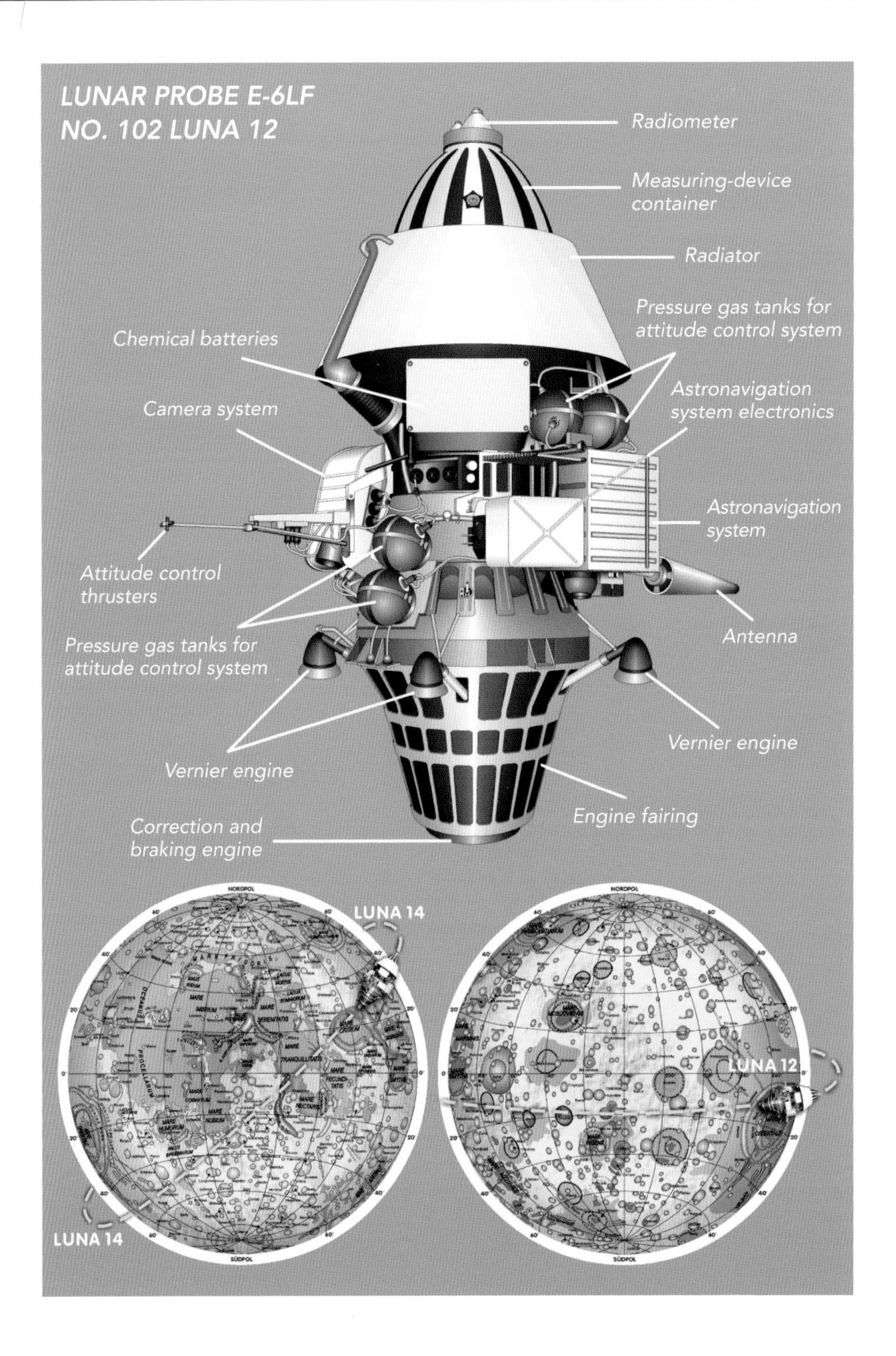
LUNAR PROBE E-6LF
NO. 102 LUNA 12
Radiometer
Measuring-device container
Radiator
Pressure gas tanks for attitude control system
Chemical batteries
Astronavigation system electronics
Camera system
Astronavigation system
Attitude control thrusters
Antenna
Pressure gas tanks for attitude control system
Vernier engine
Vernier engine
Engine fairing
Correction and braking engine
LUNA 14
LUNA 14
LUNA 12
NORDPOL
SÜDPOL
NORDPOL
SÜDPOL

of the crew transfer that would take place in lunar orbit during the later 7K-LOK flights.

The launch of the very first and unmanned Soyuz took place on November 28. It was a trouble-free launch, and all of the vehicle's antennas and solar panels deployed immediately after it entered orbit. TASS subsequently reported the launch of an Earth satellite with the designation Cosmos 133, giving no clue as to the launch's association with the manned spaceflight program. But even before the spacecraft left the area of radio contact over the Soviet Union, it became apparent that immediately after separation of the upper stage of the Soyuz rocket, the operating pressure in the pressure tank of the primary attitude control system had fallen from 340 to 38 atmospheres within 120 seconds. Fifteen minutes later the primary system's fuel was used up. No one knew what had caused it to disappear. The spacecraft was left rotating about its longitudinal axis at a rate of two rotations per minute.

This was a serious problem, since without attitude control Cosmos 133 could not be brought into the proper position in space for reentry. As well, the main engine, which was on the same circuit, could not be activated to undertake the reentry burn.

The decision now had to be made as to whether to deliver the second Soyuz to the launchpad and prepare it for launch, but first, definitive information was needed about the fate of Cosmos 133. After its first orbit, the Soyuz again came within range of the big tracking stations in Crimea, on the Caspian Sea and at Baikonur. In those days the controllers had no monitors and computers that could make all the data immediately visible to everyone. Curves and diagrams were plotted on rolls of paper, and telemetry officers brought them to specialists who, hunched over them, made their assessments. The facts soon became clear: the combined mission with the second Soyuz had to be called off.

The backup system was activated, but immediately the next problem revealed itself. The transmitted control commands resulted in changes of direction that were exactly the opposite of what was desired. That meant that the backup system was also not in a position to put the ship in the correct position in space for reentry. All that was left was one set of very weak thrusters, which were used to stabilize the spacecraft during longer drift phases, but they were not designed to keep the 15-ton spacecraft stable and in position during very dynamic maneuvers such as the braking ignition for reentry. The controllers nevertheless attempted, in combination with the onboard camera system, to use the thrusters to position the spacecraft: it would not be a precise reentry.

The maneuvers were complicated, there were many difficulties, and several times the controllers were forced to start over. But in the end it seemed to work. Then, however, the paranoia of the Cold War struck again. In the Soyuz spacecraft was a self-destruct mechanism whose purpose was to destroy the spacecraft should it go off course during recovery, to prevent it falling into unauthorized hands. This system was installed in all unmanned Soviet spacecraft, and now it prevented the spacecraft from being brought back. When the Soyuz overshot the planned landing area and approached Chinese territory, the 50-pound charge of TNT detonated.

Several weeks later, after the evaluation of all the data, it was relatively clear what had happened. The mysterious disappearance of the primary system's fuel could not be fully explained, but it appeared that a mechanical defect had occurred when the spacecraft separated from the third stage of the booster rocket. As far as the reserve control system's false control signals were concerned, an error was discovered that had not been noticed during the ground tests because the individual components had been checked separately. Just prior to the delivery of the first flight unit, the mounting of the secondary system's attitude control thrusters had been changed by 180 degrees. During the tests it was realized that their exhaust gases might damage the deployed solar panels, but because of the enormous time pressure, they forgot to also change the command switches to these thrusters. Testing each of the subsystems on their own worked perfectly, but because of the incorrect switching, in actual use the roll commands were given in the opposite direction.

Because all the other systems had worked well, and it was found that a cosmonaut onboard could have manually overridden the fault, in the end the flight was regarded as at least partially successful. The fact that they had been unable to practice the landing appeared of rather secondary importance.

The double mission now could not be carried out. It was therefore decided to send the second Soyuz up on an individual test flight to test all the systems, but especially the changes that had been undertaken after the experience gained with the first Soyuz. This launch was scheduled for December 14, 1966. At that time, everything was again ready. Telemetry operators waited on the tracking vessels *Chazma* and *Chumikan* in the Pacific, as did reception technicians on the *Dolinsk* in the Gulf of Guinea. The latter was in a fierce storm on that day; nevertheless it was still functional.

The launch was scheduled for exactly 14:00 Moscow time. Until then the preparations and countdown had gone according to plan. Then came the moment when the engines ignited. The activation of a cartridge in each of the five engines initiated the burn process. Ignition in the central stage and three of the four boosters took place normally, but the igniter in the fourth failed. The launch process was halted automatically, and the four engines that were running were stopped. By itself it was nothing extraordinary, although time consuming. After the engines had been ignited, a new launch attempt could not be made until all the combustion chambers had been inspected, all igniters had been replaced, and the cause of the fault had been determined with certainty.

After a while the launch crew was sent out to secure the rocket and inspect the engines. The service platform was swung back to the rocket to provide access to it. The crew was already at work when, with a roar, the Soyuz capsule's rescue system came to life and dragged the payload fairing, the landing capsule, and the attached orbital module into the sky with it. Before the launch crew knew it, the capsule, which was three-tenths of a mile away from the launchpad, was hanging beneath its parachute at a height of several hundred feet, floating slowly toward the steppe. Immediately afterward the two halves of the payload fairing crashed to the ground near the launch tower.

So far, so bad. This could still have ended not half badly, but then disaster struck. The solid-fuel rocket of the rescue system had set the service module, which was still atop the rocket, on fire. Flames now danced around it.

It was not difficult to imagine what happened when the fire reached the fuel

supplies inside the service module, still attached to the tip of the rocket. Over the loudspeaker came the command to immediately evacuate the launchpad, and the men began running for their lives. Meanwhile the fire had worked its way through the torn lines in the service module. First the coolant began to burn, then the peroxide system exploded and finally the hydrazine and nitrogen-tetroxide tanks went up, followed immediately by the entire rocket, which was completely filled with fuel.

Now the question arose: What in the world had set off the rescue system (which, by the way, had functioned perfectly)? While the situation after the aborted launch had been critical, it had in no way been particularly unusual. Aborted launches after ignition had started—but before the rocket lifted off—had happened often in the past, and there were procedures in place to deal with this. The answer was unbelievably simple, and yet so complex that no one had factored it in from the beginning.

The launch rescue system was designed for three activation modes. In fact, however, there was a fourth hidden mode. And the technicians first became aware of it after the catastrophe of December 14. It resulted from the spacecraft being cut off from its external power supply for thirty minutes. The gyroscope in the rocket's inertial navigation system was already running by then. Because of its extremely high revolutions, it took about forty minutes without power to run down. In the half hour that passed after the aborted launch, Earth's rotation had caused a deviation from the planned flight path of more than 8 degrees. A deviation of 8 degrees or more was defined by the onboard computer as the abort criterion for a mission that had already started. After the rescue system correctly determined that the capsule was still at low altitude, it fired, taking the crew module with it. The rescue system functioned perfectly, precisely as it was supposed to. No one could conceive of such an exotic scenario, and no one had thought that the rescue system's powerful solid-fuel rocket could set fire to and destroy the undamaged booster.

Launchpad 31, which was designed exclusively for Soyuz launches, was completely destroyed and had to be rebuilt. This meant that the Soyuz launches would have to take place from Launchpad 1, which was already in heavy use. It was the only other R-7 launch site at Baikonur.

At 11:16 Central European Time on December 21, 1966, the Soviet Union launched Luna 13. The probe landed in the Sea of Storms at 19:01 on December 24, and four minutes later it began transmitting panoramic images. It was the third soft landing on the moon, the second by the Soviet Union, and the last mission of the E-6 program. Compared to Luna 9, the landing unit was equipped with a series of additional devices and instruments. There was a stereo camera system and two retractable outriggers, several yards long. On each was a penetrometer for measuring the density of the lunar soil. There was also a boring mechanism that could dig 17.7 inches into the lunar soil.

On December 28, the batteries were exhausted and the spacecraft ceased operating.

1967: FIRST HALF OF THE YEAR

January 27: The crew of Apollo 1, consisting of Virgil Grissom, Edward White, and Roger Chaffee, are killed in a training accident two weeks before launch.

February 5: The US launches Lunar Orbiter 3. It takes 626 photos of potential Apollo landing sites. On October 9, 1967, the probe is intentionally crashed on the surface of the moon.

The flight test of the third unmanned Soyuz began on February 7, 1967. After the spacecraft reached orbit and its existence could no longer be kept secret, it was named Cosmos 140. The first test points to be completed concerned communications, the functioning of the star sensor system, and the power supply system.

All looked very good until the third orbit. Then the difficulties began. The sensors were obviously not capable of positioning the spacecraft so that the solar panels could deliver power. Moreover, in its failed attempts to align itself with the sun, the Soyuz used large quantities of attitude thruster fuel. By the fourth orbit of Earth, half of it had been used up. Once again, it was now just a case of returning the vehicle safely.

The massive consumption of fuel was finally stemmed. Controllers thus gained time to work out a return strategy. One thing was for sure: it had to happen quickly, since the batteries onboard would be dead after two days at most. The controllers finally succeeded in initiating reentry. The retrorockets fired normally. The orbital module and the service unit separated from the crew cabin, and the still fully charged landing batteries could be used for the short rest of the mission.

But then another problem arose. Along the calculated descent path the recovery forces received no signals from the capsule. It failed to appear at the calculated landing point; it had disappeared without a trace. Finally, a very weak signal was received from the ice of the Aral Sea. The capsule had come down more than 300 miles away in the frozen sea. The hot heat shield had melted through the ice, and the capsule had sunk to a depth of 30 feet. The landing parachute lay over the hole melted into the ice.

It was only after the capsule was finally recovered that the engineers got the real shock. Examination of the capsule revealed that the heat shield had burned through in several places at one spot in the upper area of the capsule. A crew without spacesuits

Mission Data, Luna 13	
Mission designation	Luna 13
Date	December 11, 1966, 11:17 CET
Spacecraft	E-6M no. 205
Booster rocket	8K78M (U103-45) Molniya M
Spacecraft weight	3,571 pounds (249 pounds landing weight)
Planned mission objective	Soft landing
Mission results	Soft landing at 19:04 on December 24, 1966, in the Mare Procellarum
Last contact	07:13 CET on December 28, 1966
Current location	18.52 degrees north and 62.3 degrees west (between the Krafft and Seleucus craters)

Launch of a Proton K with a type 7K-L1 unmanned circumlunar spacecraft.

would not have survived. Even with spacesuits, survival would have been questionable.

On March 10, 1967, a test flight with the mockup no. 2P (*P* for *prosteishiy*, which roughly translated means "simplified") of the circumlunar probe lunar vehicle on the fifth Proton K. The spacecraft had no heat shield, and thus recovery was not anticipated. TASS designated the satellite Cosmos 146; the flight was successful. The primary purpose of the mission was to test the Block D stage, which was to be used both for the circumlunar program as well as the moon-landing flights. After a day in a low Earth orbit, following reignition of Block D the combination was sent into a highly elliptical orbit. This was followed by a simulated reentry into Earth's atmosphere with a velocity equivalent to that of a return from the moon. The mission was a complete success.

Another test of the Zond complex took place on April 8, 1967. Onboard the Proton K booster rocket this time was mockup no. 3P, the last fully equipped 7K-L1 spacecraft. The desired mission profile was exactly the same as that of Cosmos 146 a month earlier. Once again, a soft landing on Earth was neither envisaged nor possible. In contrast to the flight of March 10, insertion into the lunar transfer trajectory would not take place within ninety minutes; instead, not until after twenty-four hours. The insertion into Earth orbit was successful, but the reignition of the Block D stage twenty-four hours later failed.

A later evaluation of the telemetry revealed that the cause had again been faulty wiring in the control system. The pods for the pre-acceleration engines, which were required for the second ignition, were jettisoned prematurely. Ignition could therefore not take place, and the complex was stranded in a low Earth orbit. In a brief statement, TASS called the satellite Cosmos 154. After two days in orbit the spacecraft burned up in Earth's atmosphere.

April 17: Surveyor 3 lands in the Sea of Storms. Although the probe was active for only one lunar day, it sent 6,326 photographs back to Earth. The last contact with the spacecraft was on May 7.

On March 25, 1967, the chairman of the Council of Ministers, Leonid Smirnov, held a meeting in Moscow during which a presentation was made on the progress of preparations for the manned Soyuz flights. At this meeting, Vassily Mishkin declared that the manned double flight, which would officially sound the bell for the Soyuz program, was imminent. The active vehicle, with the designation Soyuz 1, would be launched on April 21 or 22, while the passive Soyuz 2 would be launched the day after. In fact, in the end it was at 03:35 Moscow time on April 24.

Mishkin stated that the active vehicle would be manned by a single cosmonaut; the passive ship, by three crewmen. After a successful docking, two cosmonauts would leave the passive ship and make a spacewalk to the active craft. A day later both vehicles were to return to Earth. Kerim Kerimov, the state secretary responsible for spaceflight matters in the ministry for general machinery construction, confirmed that preparations were going according to plan and that there were no doubts as to the reliability of the spacecraft. The primary crew consisted of Vladimir Komarov for the active Soyuz spacecraft and cosmonauts Bykovsky, Khrunov, and Yeliseyev for the passive ship. The backup crew consisted of Yuri Gagarin for the active ship and Adrian Nikolayev, Viktor Garbatko, and Valery Kubasov for the passive vehicle.

On April 23, 1967, the TASS news agency announced the launch of the first unit of a spacecraft with the name Soyuz. Preparations had gone perfectly, and the launch itself was faultless. Immediately after the Soyuz reached orbit, however, things began to go wrong. Fault after fault accumulated, and Komarov mastered them all, but in the end, both spacecraft—Soyuz 1 and Soyuz 2—had a fatal flaw in their design that left the crew no chance. It was a fault that had gone undiscovered in the earlier tests, since it was limited exclusively to production units 4 and 5, precisely those assigned to the double mission.

Immediately after launch the left solar panel failed to deploy. In itself, it was not particularly serious matter, but the closed solar panel prevented deployment of the backup telemetry antenna, and even worse, it blocked the star sensor for the automatic alignment of the solar panels toward the sun. This prevented all automatic-navigation maneuvers, since without the sensors a manual alignment was simply not possible because the cosmonauts did not have the access parameters on the ship. The batteries would be completely discharged within two days, and there was no possibility of recharging them. Thus it was clear. There would be no rendezvous maneuver. The priority now was to return the cosmonauts safely to Earth.

The decision making was difficult. Unlike in the US, where there was a control center in which a single responsible flight controller

made all decisions, at that time there were several control centers in the USSR. One was operated by the state committee in Moscow, another was in the orbital tracking station in Crimea, and the third was in the OKB-1's facility in Baikonur. Decision making in the Moscow center proved particularly difficult. For hours, the people there could not make up their minds to abort the flight. Not until after the fifth orbit, in the eighth hour of the mission, did they acknowledge that the mission as originally planned had failed. Its problem was that it was reluctant to deliver any bad news to the leadership in Moscow, and it hoped that the mission might somehow be salvaged.

Finally, however, it was official, and the controllers began preparing Komarov's return to Earth. In the meantime the cosmonaut made every effort to position the remaining solar panel so that the batteries could be recharged, but he was unsuccessful. The energy left onboard was sufficient until the seventeenth orbit and no longer. With the backup battery he might make it to the nineteenth orbit, but definitely no longer. During the long hours in which he was out of radio contact with the Soviet flight controller, Komarov kept trying to align the spacecraft with the sun, but all his efforts failed. During the sixteenth orbit, Gagarin sent him the recovery instructions. In the short time available, however, Komarov was unable to stabilize the spacecraft so that reentry could begin during the seventeenth orbit.

The only remaining option was the nineteenth orbit (orbit 18 was out of consideration due to reasons of celestial mechanics and because new instructions had to be sent to the cosmonaut). In a desperate maneuver, Komarov was supposed to fly "by sight" and align the spacecraft with Earth's horizon. As soon as he entered the dark side, he was supposed to transfer control to the gyroscopic system, which, it was hoped, would maintain the spacecraft's position in space during this time. The decisive reentry maneuver would then take place immediately after Komarov again reached the daylight zone. There were many, many maneuvers that had to be executed precisely to the second.

The last report from Komarov after the retro burn maneuver, and immediately before the service and orbital modules were separated, was "The engine worked for 146 seconds and shut down at 5 hours, 59 minutes and 38.5 seconds." At six hours, fourteen minutes, and nine seconds came the command "Emergency 2." These were the last words that made their way through the static noise. "Emergency 2" meant that a ballistic descent would take place instead of a controlled recovery. This was due to the unstable behavior of the spacecraft during the improvised retro ignition, and it was not supposed to be a major cause for concern. It meant a significantly higher g-load for the cosmonaut, but nothing life threatening.

The recovery units under General Kastutin finally reported that a parachute had been sighted and that a landing time of six hours and twenty-two minutes was expected. That was the last report for quite a while. The flight controllers, who had been working for twenty-six hours, left the control room in good spirits and went to have an early breakfast.

The terrible news was not received until much later. The drogue chute was the only one to deploy. Its job would have been to initially slow the spacecraft and then seventeen seconds later pull the main parachute from its container. As a result of a production error, however, the inside of the container was so rough that the force exerted by the drogue chute was inadequate to pull the main parachute from its container. The backup system also failed to deploy, because the open drogue chute so disrupted the airflow on the outside of the capsule that this parachute was also unable to open. As a result, the capsule struck the Earth at approximately 93 miles per hour. The peroxide of the attitude control system caused an explosion on impact, and Komarov was killed immediately.

Flight devices 4 and 5 were built to the same criteria. If Soyuz 1's second solar panel had deployed correctly, the sun sensor and the antennas would also have deployed properly, then Soyuz 2 would have launched with exactly the same fault. Bykovsky, Khrunov, and Yeliseyev would also have lost their lives.

May 4: The US launched Lunar Orbiter 4. It took 536 photographs of potential Apollo landing sites. The probe crashed due to natural phenomena (mascons) on October 31, 1967.

On May 17, 1967, the Soviet Union launched a research probe whose special task it was to research the mass concentrations on the moon discovered by Luna 10 and to test the telecommunications equipment of future Zond vehicles. TASS designated the spacecraft Cosmos 159. The source here is not completely certain. It is probable that no lunar orbit was planned, instead just a highly elliptical Earth orbit that passed near the moon. It is also not completely certain if Cosmos 159 was a type E-6LS space probe. This mission is listed as a success in the R-7 launch statistics. Instead of the probable target apogee of 155,000 miles, the spacecraft reached an altitude of only 37,655 miles. The perigee, 120 miles, was so low that Cosmos 159 crashed back into Earth's atmosphere on November 11, 1967.

1967: SECOND HALF OF THE YEAR

July 14: Contact with Surveyor 4 was lost just 150 seconds before it touched down on the moon. The abrupt loss of radio contact is suggestive of an explosive event toward the end of the burn phase of the probe's solid-fuel descent engine.

July 19: The Americans launched Explorer 35 into a high lunar orbit. The spacecraft delivered physical data from the interplanetary area in lunar distance. The probe remained in operation until June 24, 1973, and was then shut down. Its fate is unknown, but because of lunar orbit interference it must have crashed onto the surface of the moon in the late 1970s.

Mission Data, Cosmos 159	
Mission designation	Cosmos 159
Date	May 16, 1967, 22:43 CET
Spacecraft	E-6LS no. 111
Booster rocket	8K78 (Ya716-58) Molniya
Spacecraft weight	approx. 3,300 pounds
Planned mission objective	Test the communications and data transmission system for the Zond program. Circumlunar flight
Mission results	Premature burnout. Reached a perigee of only about 120 miles and an apogee of 37,665 miles
Fate	Entered Earth's atmosphere on November 11, 1967

August 1: The US launched Lunar Orbiter 5, which subsequently took 844 photos of Apollo landing regions. The spacecraft was intentionally crashed on January 31, 1968.

September 8: Surveyor 5 was launched to the Sea of Tranquility and from there transmitted more than 19,000 photos of the landing site until December 17.

A 7K-L1S Zond spacecraft is prepared for launch.

For weight reasons, the type 7K-L1 circumlunar spacecraft had been "freed" of an entire series of safety-related systems. There was no reserve landing parachute, and it carried just 880 pounds of fuel for attitude control and course changes. In any case, the small spacecraft was powered by the extremely reliable Isayev AK-27 engine. The two solar generators were now each equipped with three panels instead of four as on the standard Soyuz 7K-OK, and the crew of two had to spend the seven-day trip around the moon in a spatial volume of just 88.3 cubic feet, since the orbital module of the 7K-OK had to be dispensed with for weight reasons.

A perfect Zond mission would take place roughly as follows:

First would be the launch of the three-stage Proton K, which was supposed to put itself and the orbital unit, consisting of Block D and the 7K-L1 Zond spacecraft, into a near-orbit suborbital trajectory. Immediately afterward, after a free-flight phase of several minutes, followed the first ignition of the only partly fueled Block D stage to transport the orbital unit into a low parking orbit. It was partly fueled because a fully fueled Block D stage would have exceeded the Proton K's payload limit. Remember: the Block D stage was optimized for the N1-L3 program and not for the L1 program. The transition orbit was supposed to have a perigee of 118 miles, an apogee of 137 miles, and an inclination of 51.6 degrees to the equator.

The two cosmonauts onboard would now check the spacecraft's systems and jettison the support structure for the payload fairing (one of the many peculiarities of the 7K-L1). Ignition of Block D for the translunar acceleration maneuver was supposed to take place in the period between ninety minutes after achieving Earth orbit and one full day afterward.

Then Block D was jettisoned. The cosmonauts then had to place the spacecraft in a sun-oriented, so-called "barbecue" mode, in which it turned slowly about its longitudinal axis. It was supposed to complete one full rotation every six minutes. This would have been necessary to avoid thermal overloading of various systems. One or two course corrections would have been carried out with the Isayev S5.53 engine en route to the moon.

The flight around the moon at a distance between 621 and 7,456 miles would then follow on the third day. The two cosmonauts would take photographs, shoot film, and conduct scientific experiments in proximity to the moon.

During the return flight to Earth—we find ourselves on a free return trajectory—one or two more burn maneuvers would have been necessary by the main engine.

Prior to reentry into Earth's atmosphere, on the seventh day of the flight, first the large parabolic antenna and then the 7K-L1's instrument section would have been jettisoned. Then reentry into Earth's atmosphere would begin. This was supposed to take place using the "skip-reentry" method, which placed the lowest g-loads on the crew. If this maneuver failed, then a hard ballistic descent would be necessary. Under nominal conditions the landing would take place in Kazakhstan, not far from the launch site.

That was the plan. The first mission by a fully equipped Zond, designated 7K-L1 no. 4L, began on September 27, 1968, but the launch failed during the first phase of flight by the Proton K booster rocket. A sealing part got into one of the turbopumps, causing one of the first-stage engines to cut out seconds after launch. The rocket continued to climb slowly but subsequently lost its stabilization, and in the ninety-seventh second of flight the automatic command was given to shut down all engines. The launch rescue system brought the capsule safely to the ground 33 miles from the launch site. Zond 7K-L1 no. 4L was the first of a total of nine fully equipped 7K-L1

Mission Data, 7K-L1 No. 4	
Mission designation	7K-L1 no. 4
Date	September 28, 1967, 22:43 CET
Spacecraft	Soyuz 7K-L1 no. 4
Booster rocket	Proton K 8K78K
Spacecraft weight	11,383 pounds
Planned mission objective	Circumlunar flight. Test of the Zond spacecraft
Mission results	One of the first-stage engines shut down immediately after launch. The rocket's flight became unstable, and the other engines were also shut down
Fate	The spacecraft was separated from the rocket by the rescue system and landed 35 miles from the launch site

units that would be launched in the next two and a half years.

After Korolev's death, the state committee decided to continue the manned Soyuz flights until all of the spacecraft's defects had been addressed, especially problems with the landing system. The next two flight units, 7K-OK nos. 5 and 6, were subsequently prepared for an unmanned mission in October. Summer passed with tests of the landing system, and dozens of drops of dummy spacecraft from helicopters and aircraft. The other systems also underwent rigorous trials on the test stands, which revealed many more previously concealed faults.

At 12:30 Moscow time on October 27, Soyuz spacecraft 7K-OK no. 5, with the call sign Amur, was launched from Launch Complex 31 at Baikonur. TASS gave it the designation Cosmos 186. The device was checked in orbit for about three days, then another series of malfunctions appeared. Twice, orbit-raising maneuvers failed on the initial attempt, and there were again difficulties with the sun sensor, but the controller on the ground was able to handle all the problems.

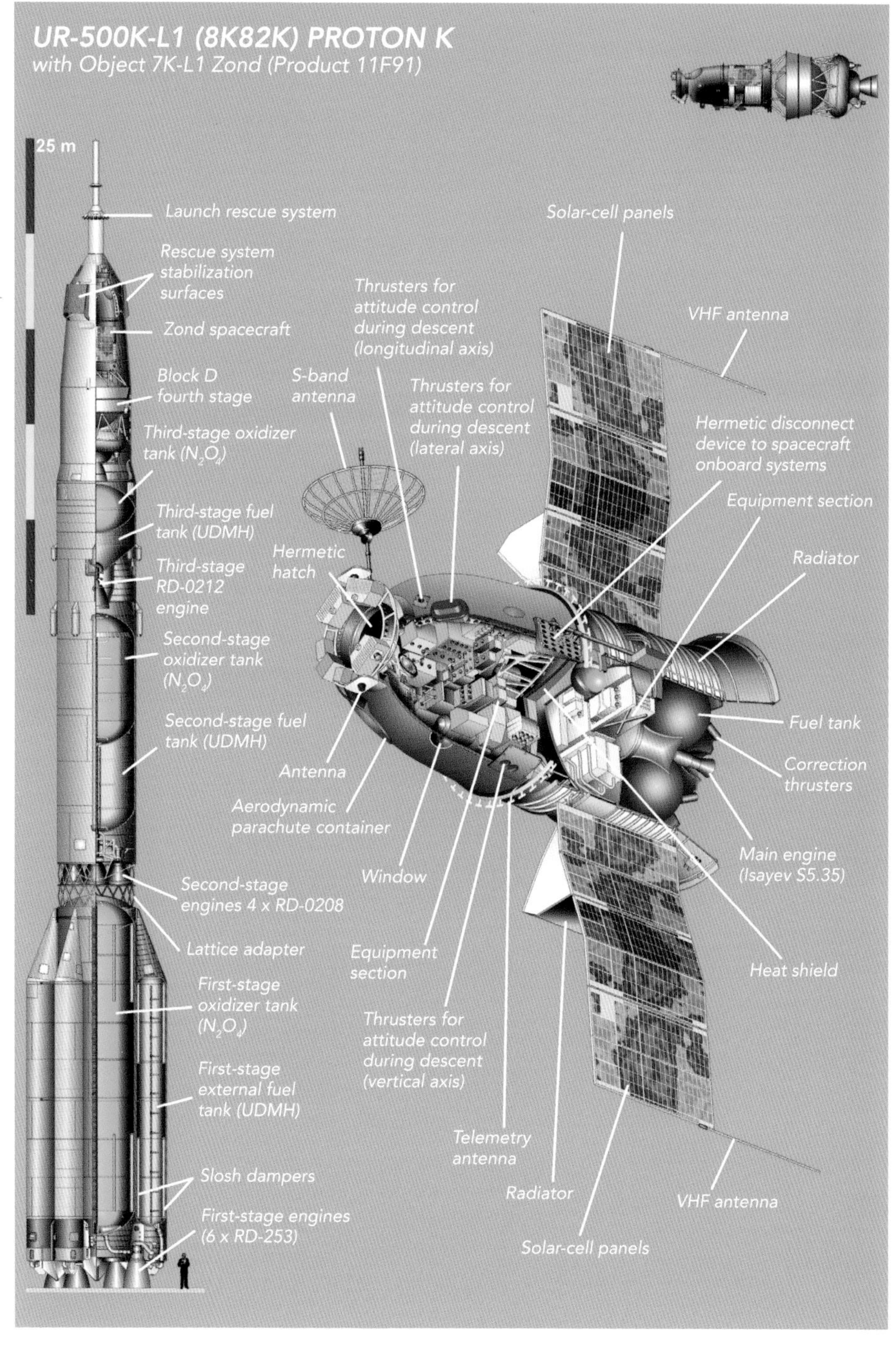

UR-500K-L1 (8K82K) PROTON K
with Object 7K-L1 Zond (Product 11F91)
25 m
Launch rescue system
Rescue system stabilization surfaces
Zond spacecraft
Block D fourth stage
Third-stage oxidizer tank (N_2O_4)
Third-stage fuel tank (UDMH)
Third-stage RD-0212 engine
Second-stage oxidizer tank (N_2O_4)
Second-stage fuel tank (UDMH)
Second-stage engines 4 x RD-0208
Lattice adapter
First-stage oxidizer tank (N_2O_4)
First-stage external fuel tank (UDMH)
Slosh dampers
First-stage engines (6 x RD-253)
Solar-cell panels
VHF antenna
Thrusters for attitude control during descent (longitudinal axis)
S-band antenna
Thrusters for attitude control during descent (lateral axis)
Hermetic disconnect device to spacecraft onboard systems
Equipment section
Radiator
Hermetic hatch
Fuel tank
Correction thrusters
Antenna
Aerodynamic parachute container
Main engine (Isayev S5.35)
Window
Equipment section
Heat shield
Thrusters for attitude control during descent (vertical axis)
Telemetry antenna
Radiator
VHF antenna
Solar-cell panels

At 11:13 Moscow time on October 30, 7K-OK no. 6, with the radio call sign Baikal, followed into orbit and was given the official designation Cosmos 186 by TASS. Two days later it docked automatically and outside the area of influence of the control center with Cosmos 188. There were problems with the rendezvous maneuver and docking, however. The automatic system had used an excessive amount of fuel for the approach. A total of no fewer than forty-five correction burn maneuvers took place, and in the end, only what is now known as a "soft docking" took place. The locking clamps engaged, but the electrical connections were not made. The two spacecraft were still 3.5 inches apart. Something was preventing the spindle drive from bringing the two spacecraft completely together and activating the plug connections.

Flying in this "loose" connection was dangerous, and so it was decided to separate the two spacecraft after two orbits of Earth and return them to Earth as soon as possible. The sun sensor on 7K-OK no. 5 (Amur) again failed, resulting in a ballistic descent with high g-loads. The flight controller subsequently left Baikal in orbit for another day to determine what had caused the failure of Amur's sun sensor. This did not help much, since Cosmos 188 made such an imprecise—in this case, too flat—descent that the self-destruct blew up the spacecraft since it was feared it might come down in China. The subsequent assessment of the descent trajectory revealed that had the capsule not been destroyed, it would have come down near the city of Chita, 240 miles from the Chinese border.

As a precondition for manned flights by the Zond, the responsible government committee decided that initially only four pure experimental flights should be undertaken. These would have to be followed by at least two successive, completely successful unmanned operational flights. Only then would the 7K-L1 be released for manned missions. Use of the unpredictable Proton K meant that the risk during launch was very high, even though this could be offset with the very reliable launch rescue system. But the limitations of the capsule system itself, in which many redundancies had been eliminated, made a manned flight a risky game even under the most-favorable conditions.

On November 22, 1967, the second launch of an unmanned circumlunar Zond spacecraft also failed during the initial launch phase. This time one of the four second-stage engines of the Proton K booster rocket failed to ignite. This incident activated the rescue system, which delivered the capsule to safety. Despite the failed launch, once again the crew would have been saved; however, the landing after the aborted launch was unusually rough, because Zond 7K-1L no. 5L's braking rocket ignited high in the air and not at the moment of touchdown. As well, there was a very strong wind at the emergency landing site that dragged the capsule 1,500 feet through the terrain after landing.

Mission Data, 7K-L1 No. 5	
Mission designation	7K-L1 no. 5
Date	November 22, 1967, 20:08 CET
Spacecraft	Soyuz 7K-L1 no. 5
Booster rocket	Proton K 8K78K
Spacecraft weight	11,383 pounds
Planned mission objective	Circumlunar flight. Test of the Zond spacecraft
Mission results	One of the second-stage engines failed to ignite, whereupon the three other engines were shut down
Fate	The spacecraft was separated from the rocket by the rescue system and landed 177 miles from the launch site

1968: FIRST HALF OF THE YEAR

January 9: Surveyor 7 lands in the crater Tycho and by February 21 transmits more than 21,000 photographs.

The mission of February 7, 1968, is numbered among the numerous failed launches that took place during the Soviet moon flight program. On that day the second unit in a series of three flights was supposed to be launched, with which a lunar orbiter of the E-6LS series was to prepare the manned moon landings. The purpose of the mission was to photograph planned landing sites and measure mass concentrations beneath the lunar seas. But once again, the booster rocket failed and the spacecraft failed even to reach a parking orbit around Earth. The fault this time: a fuel valve stuck in the "full open" position. This caused the gas generator to be starved of fuel from the 525th second of flight onward, quite simply because the open valve caused it to be used up prematurely. The engine subsequently shut down.

At the beginning of 1968, eleven cosmonauts were already at an advanced stage of training for the manned Zond flights. Each mission would be carried out by an experienced cosmonaut as commander and a space rookie. The most-prominent members of this corps of cosmonauts were Alexey Leonov and Pavel Popovich.

On the evening of March 2, 1968, a Proton K was launched with Zond 4. Despite this designation, it was in fact the third Zond

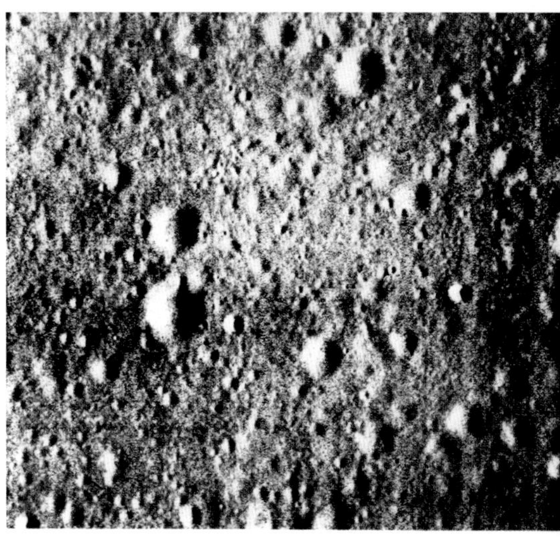

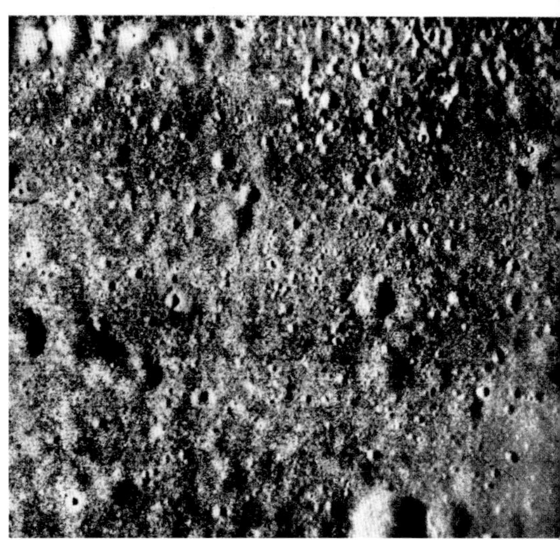

Photos of the surface of the moon taken by Luna 12. They were made at an altitude of 155 miles and show a region south of the crater Aristarch. Photos of similar quality were also expected from photo orbiter E-6LS no. 112.

Mission Data, E-6LS No. 112	
Mission designation	—
Date	February 7, 1967, 10:44 CET
Spacecraft	E-6LS no. 112
Booster rocket	8K78M (Ya716-57) Molniya M
Spacecraft weight	3,748 pounds
Planned mission objective	Lunar orbiter
Mission results	Failed to reach Earth orbit because of a valve failure in the booster's third stage.
Fate	Crashed back to Earth after about 30 minutes.

mission with the fully equipped 7K-L1 spacecraft. The TASS news agency had simply assigned it the next number after Zond 3, which had been launched on July 18, 1965, for a flypast of the moon. If one also counts the flights with mockups, and the failed missions in the early launch phase (about which the world never learned anything), then it was in fact the sixth Zond.

The launch and insertion into orbit, with the first ignition of Block D, went perfectly. One hour and eleven minutes after leaving the launchpad, Block D fired a second time. Burn duration was 459 seconds, and this placed Zond 4 on a highly elliptical Earth orbit with an apogee of 220,000 miles. A flight around the moon was not planned for this mission. To precisely determine the spacecraft's characteristics, they even launched in the opposite direction of the moon, since its gravity would have interfered with the specific tasks of this flight. Zond 4 was the first Soviet spacecraft equipped with an onboard computer, the Argon 11 unit, which weighed 75 pounds.

The first problems soon arose. The omnidirectional antenna had obviously not fully deployed, and the 100K star sensor caused problems from the outset because it was unable to find the reference stars Sirius and Canopus. The reason was a stream of hydrogen-peroxide particles flowing out of the attitude control system, which sparkled in the sunlight and irritated the sensor. As a result, spatial orientation for the course correction maneuver was not possible.

Consideration was now given to resetting the calculation mode so that the bright planet Venus could be used as a new reference point. While this was made to work after several attempts, it was rejected as too dangerous because Venus was too near the sun, and a direct "look" into the sun would ruin the sensor. Finally, Sirius was retained as the reference star, and a complex procedure was worked out with which the sensor's sensitivity could be adapted to the brightness of Sirius so as to ignore the peroxide particles. The probe lost its reference star several times, and new calibrations had to be carried out, no easy matter over a distance of 180,000 miles.

Finally, the necessary course corrections were made. On March 6, the probe passed the apogee, and its fallback toward Earth began. To reach the predetermined corridor, another course correction was necessary at 96,000 miles from Earth on the morning of March 9. Landing was supposed to take place at about 21:00 Moscow time the same day.

For the reentry, a flight path was again chosen that in American space jargon was called a "skip reentry." It was chosen to minimize g-forces for a crew during a return at near-Earth escape velocity. The spacecraft entered Earth's atmosphere at a shallow angle, briefly left it again, and finally entered again to achieve the final landing approach. One can compare this process to skipping a stone on the surface of the water. The stone loses its energy by repeatedly bouncing on the water before finally sinking.

Problems with the star sensor reappeared shortly before reentry into Earth's atmosphere. The spacecraft was no longer able to maintain its exact position in space for the controlled descent and went into a ballistic descent—the programmed emergency mode. This resulted in pressure loads of 20 Gs. Surviving this would have been extremely problematic for a crew. With the transfer to the ballistic mode, the envisaged target area in Kazakhstan was no longer achievable, and the capsule would come down over Africa, the Mediterranean, or Turkey. This automatically activated the capsule's self-destruct mechanism, and the Zond 4 landing capsule was destroyed.

It was later discovered that the second deviation by the star sensor had not been caused by escaping peroxide like the first interruption. It was instead due to black paint that came off the interior of the sensor housing, and the drifting particles irritated the sensor.

Yuri Gagarin lost his life on March 27, 1968. This was a huge blow to the Soviet Union. Gagarin had just succeeded in having himself returned to active flight duty. For a long time he had been presented solely as a national icon. When he was killed in a MiG-15 during a training flight, he was already envisaged as the standby for the Soyuz docking mission and was also a candidate for the circumlunar L1 program. The accident occurred under conditions that have still not been fully explained.

The lunar orbit mission of Luna 14 began on April 7, 1968. It was the third E-6LS mission and the first one to go successfully. The two previous flights had failed during the launch phase. The probe tested the communications equipment for the planned Soviet manned moon landing. The flight was the last use of a Soviet E-6 series spacecraft, which represented the second generation of unmanned lunar research probes. Its configuration was very similar to that of Luna 12. It also meant the end of the R-7 8K78 Molniya as a booster rocket in the Soviet lunar program. From then on, spacecraft of the E-8 series were used, launched by the Proton K.

After testing of the Soyuz system by using the spacecraft Cosmos 186 and Cosmos 188 had again revealed serious problems with the system, orders were given for another experimental mission with two unmanned

Mission Data, Zond 4	
Mission designation	Zond 4
Date	March 2, 1968
Spacecraft	Soyuz 7K-L1 no. 6
Booster rocket	Proton K 8K78K
Spacecraft weight	11,331 pounds
Planned mission objective	Highly elliptical Earth orbit with an apogee of 205,000 miles. Test of Zond spacecraft.
Mission results	Planned orbit achieved. Landing failed.
Fate	Reentered Earth's atmosphere too steeply. Capsule was destroyed by self-destruct system at an altitude of 6 miles.

Mission Data, Luna 14	
Mission designation	Luna 14
Date	April 7, 1968, 11:09 CET
Spacecraft	E-6LS no. 113
Booster rocket	8K78M (Ya716-58) Molniya M
Spacecraft weight	3,748 pounds
Planned mission objective	Lunar orbiter. Test of communications equipment for the manned N1-L3 flights.
Mission results	Lunar orbiter. On April 10, reached an orbit of 99 x 540 miles with an inclination of 42 degrees.
Last contact	Not known
Current location	Not known

spacecraft. The "active" vehicle, LOK unit no. 8, was scheduled for 13:00 on April 14, 1968. The "passive" ship, LOK unit no. 7, was supposed to be launched the next day. The precise launch time could not be determined until the orbit of the active vehicle had been precisely measured.

The launch on April 14 went perfectly, and 531 seconds after leaving the launchpad, spacecraft no. 8 reached orbit. Immediately afterward, all the antennas and solar generators were deployed. During the second orbit the TASS news agency reported that the Soviet Union had launched the spacecraft Cosmos 212.

The first course corrections proved trouble free. The new 76K sensor, which had replaced the unreliable 75K unit, worked perfectly. There was just a problem with overcharging of the silver-zinc batteries. The time for the launch of the second flight unit was already approaching when, two hours before its launch, the alarm signal was received that for unknown reasons the performance of the roll control thruster was at just 10 percent.

There now followed a hectic analysis of the problem. The second Soyuz rocket was on the launchpad, fully fueled. Postponing the launch would possibly mean calling off the entire rendezvous program, because another launch attempt the following day was not possible for technical reasons. Although the problem with the roll control thrusters could not be solved by launch time, those responsible gave the launch order for spacecraft no. 7. It was a risky decision, since if the problem with Cosmos 212 could not be solved quickly, then the rendezvous and docking maneuver would fail. There wasn't much time, since the maneuver was supposed to begin immediately after the launch of the second unit.

However, Cosmos 213, the name TASS gave the second spacecraft after launch, docked with Cosmos 212 without problem. The mysterious behavior of Cosmos 212's roll control thrusters had disappeared as mysteriously as it had begun. This time a "hard docking" was achieved with the establishment of a complete connection between the two spacecraft, including power and data links. The two vehicles remained coupled together for three hours and fifty minutes before they received the command to separate.

But the mission did not end without problems. With Cosmos 212 in its fifty-first orbit, during an orbital correction maneuver the automatically controlled orbit correction and landing engine failed for initially unexplained reasons. It quickly turned out that this had happened because after the previous flights, a new safety system had been installed that prevented the engine from igniting if the vehicle's position in space was incorrect. But now the old system functioned perfectly, and the unit, which was actually supposed to have improved the security of the entire system, failed.

What happened next was one of the still largely unknown heroic efforts of the Soviet space program. With the help of an onboard television camera, the landing procedure was reprogrammed with the help of data from the gyroscopic system, and the spacecraft was flown almost as if there were a cosmonaut onboard flying the spacecraft manually.

The maneuver succeeded. The ignition timing and thus the reaching of the landing zone were understandably imprecise. For a time it was feared that the self-destruct mechanism had been activated, but the landing was successful. The capsule did not even go into the passive ballistic mode, instead remaining actively controlled. Only during touchdown were there problems. For safety reasons the charges for parachute separation after landing

had been deactivated. There was a strong wind in the landing area, and the parachute dragged the capsule 3 miles through the terrain. Spacecraft number 7, whose descent from space on April 20 was initially normal and trouble free, experienced almost the same thing. Its parachute ultimately deployed, however, but static built up as it was being dragged, and the explosive charges activated.

The mission of the Zond spacecraft 7K-L1 number 7 was scheduled to begin at 02:01 Moscow time on April 23, 1968. The spacecraft was supposed to reach transfer orbit around Earth at 589 seconds after liftoff from the launchpad, but after 260 seconds the second stage's engine shut down prematurely because of a false signal from the crew rescue system. The rocket received this signal from the capsule and thus had no part in this flight abort. The 7K-L1's capsule's nervous rescue system again functioned extremely well, but the mission was nevertheless ruined. The capsule landed intact 310 miles from launch point. This was the third failed launch of a Proton K that was part of the Zond program in which the launch rescue system could be tested under operational conditions. Thus, the positive aspects if the failed mission ran out.

Mission Data, 7K-L1 No. 7	
Mission designation	Zond
Date	April 22, 1968
Spacecraft	Soyuz 7K-L1 no. 7
Booster rocket	Proton K 8K78K
Spacecraft weight	11,331 pounds
Planned mission objective	Circumlunar flight. Test of Zond spacecraft
Mission results	A short circuit in the flight control system of the Proton's second stage shut down the engines 260 seconds after liftoff
Fate	Spacecraft was separated from the rocket by the rescue system and landed about 323 miles from the launch site

1968: SECOND HALF OF THE YEAR

On July 15, four days before the next planned 7K-L1 Zond mission, the Block D stage and the spacecraft underwent several tests on the launchpad. The entire combination of booster rocket and orbital unit were already fully fueled when Block D's oxidizer tank suddenly exploded. One of the technicians was killed on the spot, and a second was badly injured. The accident happened because one of the telemetry lines was damaged. As a result the Block D fueling team could not realize that it had built up an overload in the tank.

The situation was extremely dangerous. Block D, with the spacecraft on top, was leaning sharply toward the rocket and was held in place only by a launchpad access arm, in which the tip of the rescue tower had become caught. The bending had caused

Block D to bore deeply into the structure of the booster rocket's third stage, and it was initially unclear if one of the Proton's fuel tanks had been damaged.

There now stood this wobbly combination that might collapse at any moment, with hundreds of tons of highly toxic hydrazine and nitrogen-tetroxide, several tons of kerosene and liquid oxygen, half a ton of solid rocket fuel in the rescue system, 53 pounds of TNT in the capsule's self-destruct system, 66 pounds of concentrated hydrogen peroxide, and 150 detonators, which supported the various pyrotechnic processes of a rocket launch that might detonate at any minute.

It required weeks of extremely dangerous work to remove the fuel from this wonky structure, and to disassemble and remove it piece by piece. The planned launch window for the Zond flights in July and August was now an illusion. Along with it, the chance of at least beating the Americans to a circumlunar flight was diminishing more and more.

Another unmanned test flight of the Soyuz 7K-OK took place on August 28, 1968. TASS named the spacecraft Cosmos 238, and the four-day mission was a success. The spacecraft was a passive variant of the Soyuz. All the changes implemented in the system after the failed mission of Soyuz 1 were tested together. The primary purpose of this mission was to build trust for the imminent resumption of manned missions.

The circumlunar mission of Zond 5 began on September 14, 1968. After the partial success of Zond 4 in March 1968, this time, and for the first time ever in the L1 program, the flight was relatively successful. After the three basic stages of the Proton had fired, there followed the usual four-minute-long free-flight phase. Then the first ignition of the Block D phase took place with a burn duration of 108 seconds. Then the parking orbit, with a perigee of 118 miles and an apogee of 136 miles, was reached. This was followed by a sixty-seven-minute free-flight phase until Block D fired a second time and placed Zond 5 on the translunar trajectory.

The only course correction en route to the moon took place on September 17, at a distance of 202,000 miles from Earth. It was sufficient to put Zond 5 precisely around Earth's satellite and then return it to Earth. Onboard were a number of organisms: two tortoises, vinegar fly eggs, mealworms, and several plants and cell samples. On September 22, during the return to Earth, the planned "skip reentry" did not succeed, and the capsule had to make a ballistic descent with high-pressure forces. The vehicle was thus unable to reach the landing area in Kazakhstan, instead coming down in the backup landing zone in the Indian Ocean. There the capsule was fished out of the water by the recovery vessels *Borovichi* and *Vassily Golovin*. When the capsule landed, the ships were a good 60 miles from Zond 5.

Mission Data, Zond 5	
Mission designation	Zond 5
Date	September 15, 1968, 22:42 CET
Spacecraft	Soyuz 7K-L1 no. 9
Booster rocket	Proton K 8K78K
Spacecraft weight	11,850 pounds
Planned mission objective	Circumlunar flight. Test of Zond spacecraft with animal test subjects onboard
Mission results	Mission successful. Flew around the moon at a distance of 1,150 miles; however, ballistic descent with high g-loads; landed in the Indian Ocean
Current location	Zond 5 can now be seen in the museum of the RKK Energia

Photo of the Earth taken by Zond 5 from a distance of 42,000 miles from Earth.

The flight caused concern in NASA. The US military had observed Zond 5 on its entire flight around the moon. During the recovery the American destroyer USS *McMorris* was nearby and photographed the entire action. The successful flight of Zond 5 was one of the instigators of NASA's decision to risk a manned lunar orbital flight with Apollo 8 in December 1968.

October 11: The United States launches the spacecraft Apollo 7. Onboard are astronauts Schirra, Eisele, and Cunningham. The eleven-day mission is a complete success.

The mission of Soyuz 3 with cosmonaut Georgi Beregovoy was scheduled for October 26, 1968. Prior to the mission there were heated discussions about the qualification of the pilot, because he had achieved poor results in the so-called "exams," the standard test prior to a flight. His alternates, Vladimir Shatalov and Boris Volynov, scored much better. After "retesting," Beregovoy was confirmed as the pilot of Soyuz 3. His task was to dock with Soyuz 2, which was supposed to reach orbit a day before him.

The launch of Soyuz 2, 7K-OK spacecraft no. 11, succeeded on October 25, 1968, precisely and as envisaged. The day after, Beregovoy followed in Soyuz 3, 7K-OK spacecraft no. 10, in an equally precise launch. His mission began on Launchpad 31, and Beregovoy became the first man to be sent into orbit from that facility. All other manned Soviet launches had taken place from installation 1/5. At forty-seven years of age, Beregovoy was the oldest person to date to go into orbit.

The planned docking with Soyuz 2 failed, however, on account of a serious mistake by Beregovoy. Only the automatic part of the rendezvous maneuver, in which Soyuz 3 was guided to 600 feet of Soyuz 2, was successful. As a result of Beregovoy's error, no docking took place. Soyuz 2 landed after three days in orbit, Soyuz 3 after four days.

After the mission of Zond 5 in September was evaluated as a total success, expectations for Zond 6 were high. If this was to be the second successful mission in a row, then the subsequent flight, planned for January 1969, could be manned. Three crews were already preparation: Crew 1 with Alexey Leonov as commander and Oleg Makarov as pilot, Crew 2 with Valery Bykovsky as commander and Nikolai Rukavishnikov as pilot, and Crew 3 with Pavel Popovich as commander and Vitaly Sevastyanov as pilot.

After the launch of Zond 6 on November 10, 1968, everything initially went completely according to plan. Insertion into the transfer trajectory and the course corrections succeeded, and on November 13, Zond 6 flew around the moon and came to within 1,615 miles of it. The automatic cameras took their color photos as planned, and the return also seemed completely normal at first. A few hours prior to landing in Kazakhstan, however, a seal failed and the air began to escape. At first the pressure stabilized at 380 millibars, but during the recovery it fell to 25 millibars and all the experimental animals onboard died. The parachute opened as planned. But then there was an electrical discharge, caused by the near vacuum in the capsule, which led to a failure of the gamma ray altimeter and initiated the touchdown sequence. The landing engine fired and the parachute, which was fully deployed by 17,400 feet, was separated by the explosive

Mission Data, Zond 6	
Mission designation	Zond 6
Date	November 10, 1968, 20:11 CET
Spacecraft	Soyuz 7K-L1 no. 12
Booster rocket	Proton K 8K78K
Spacecraft weight	11,850 pounds
Planned mission objective	Circumlunar flight. Test of Zond spacecraft with animal test subjects onboard
Mission results	Flew around the moon at a distance of 1,503 miles. Cabin decompressed before recovery, killing all the animal test subjects. The landing system separated the parachute at a height of 15,000 feet
Fate	Crashed on November 17, 10 miles from the launch site

guillotine, as it the capsule had already reached the ground. Zond 6 crashed into the ground, just 10 miles from where it had been launched six days and nineteen hours earlier.

December 21: Apollo 8 launched on its historic first manned flight around the moon with astronauts Borman, Lovell, and Anders. Unlike the Soviet Zond missions, this was not a circumlunar flight but a complete orbital mission with ten orbits of the moon. The mission was a triumphant success.

In the face of the accomplishments by the American Apollo program, sometime toward the end of 1968 the Soviets abandoned their plan to at least beat the Americans with a circumlunar flight. A launch window had opened on December 9, prior to the flight of Apollo 8, and 7K-1L spacecraft no. 13 was ready to launch, as was its Proton booster rocket. Some of the cosmonauts even sent a letter to the Politburo in which they declared their readiness to risk a manned flight. There was no response, however.

There was a meeting of the Council of Ministers on December 30, during which discussions were held as to how to respond to the American accomplishments after the sensational mission of Apollo 8. For the first time, consideration was given to halting the L1 flights, which after the manned circling of the moon by the Americans now made little sense. It was also decided to accelerate the E-8 program, which in addition to orbital lunar probes also included the return to Earth of unmanned lunar-landing probes. It was also at this meeting that the claim that the Soviet Union had never intended to carry out a manned landing on the moon and instead always planned only unmanned landings, which would remain the official version for almost two decades, was first voiced. Anything else, it was claimed to the outside world, would have been a foolish risk of human life in which they would never have taken part.

Artistic (and technically not overly accurate) representation of the return launch of Luna 16 to Earth.

1969: FIRST HALF OF THE YEAR

The first of these missions was the successful flight of Soyuz 4 and Soyuz 5. These missions finally realized what the Soviets had failed to achieve in the previous missions, including the flight of Soyuz 2 and 3. Soyuz 5 launched on January 14, with Vladimir Shatalov onboard. He was actually supposed to have gone on this trip a day earlier, but a calibration problem with the attitude control gyroscope delayed the launch. Shatalov was already in the spacecraft and engine start was only thirty minutes away when the mission was called off for a day, and Shatalov had to disembark. In itself this was not a major problem, and in favorable weather conditions the repair could have been made on the launchpad. But the outside temperature was –11 degrees F and a strong wind was blowing, and therefore the task was considered too risky.

The next day, Soyuz 5, carrying cosmonauts Boris Volynov, Yevgeny Khrunov, and Alexey Yeliseyev, was launched. On January 16, the two spacecraft docked using the automatic method. Yeliseyev and Khrunov transferred to Soyuz 4 and landed with Shatalov. Volynov came back alone in Soyuz 5.

Soyuz 4 landed on January 17, and the landing was nominal. During the landing of Soyuz 5 on January 18, however, there were considerable problems. The service module failed to separate from the capsule after retro ignition, and it did not break away until after entry into the atmosphere due to friction and heat. Thus the first part of the reentry took place with the topside of the capsule facing forward, not the heat shield. The heat shield on the topside of the capsule was much thinner than on its base. Volynov had to fear that it might completely burn away. It was also possible that the melting mass of the heat shield might cause the parachute housing doors to become stuck shut.

He experienced a ballistic descent with high g-loads. The capsule's center of gravity caused it to turn in the right direction during the descent, but there seemed to be some sort of problem—it could not be determined with certainty—with the parachute doors, since the chute came out of the compartment tangled and twisted. Fortunately it became untangled in the last few yards before striking the Earth, but the force of impact was extremely high. Miraculously Volynov was not injured. Seven years later he returned to orbit.

The four cosmonauts had an even more dangerous adventure four days later while driving in a motorcade from Vnukovo airport in Moscow to the Kremlin for the then-customary state banquet. An assassin had positioned himself directly in front of the Borovitskaya Tower in the Kremlin, traditionally the passage used by Soviet leaders, waiting for Brezhnev. He fired at the wrong vehicle, however, picking the car occupied by cosmonauts Beregovoy, Leonov, Nikolayev, and Tereshkova. The driver of their car was fatally wounded. The celebration in the Kremlin went ahead anyway, as if nothing had happened.

On January 20, 1969, another Zond mission failed in its initial phase. It was the same spacecraft (7K-L1 no. 13) and the same rocket that several cosmonauts had requested for a manned flight in December. During the second-stage burn, 313 seconds after leaving the launchpad one of the four engines shut down, about twenty-five seconds before the planned burnout. The other engines continued operating, and stage separation took place at too low an altitude and too low a speed. The third stage nevertheless ignited and again there was a malfunction, when a fuel line to the gas generator

Mission Data, 7K-L1 No. 13L	
Mission designation	—
Date	January 20, 1969, 20:11 CET
Spacecraft	Soyuz 7K-L1 no. 13L
Booster rocket	Proton K 8K78K
Spacecraft weight	11,850 pounds
Planned mission objective	Circumlunar flight. Test of Zond spacecraft
Mission results	A second-stage engine failed, and the third-stage engine experienced premature burnout. Failed to achieve orbit
Fate	The rescue system delivered the capsule to safety

Mission Data, E-8 201	
Mission designation	—
Date	February 19, 1969, 07:48 CET
Spacecraft	E-8 201 Lunokhod
Booster rocket	Proton K 8K78K
Spacecraft weight	12,324 pounds (rover 1,851 pounds)
Planned mission objective	Mobile moon exploration by the Lunokhod rover
Mission results	Proton's payload fairing collapsed after 51 seconds of flight; booster rocket exploded
Fate	Total loss of rocket and payload

broke and the engine shut down in the 500th second of flight. The rescue system then fired and took the capsule to safety in a suborbital trajectory. The soft landing by parachute occurred not far from the Mongolian border.

On February 19, 1969, another lunar mission failed during the initial launch phase. This mission was supposed to send the first Lunokhod lunar vehicle, rover no. 201, to the moon. Unfortunately, however, the Proton K failed less than a minute after leaving the launchpad. The reason for this was the special payload fairing, which was first used for the Lunokhod launch. It collapsed as the rocket flew through the zone of maximum dynamic loads. A falling piece of wreckage pierced one of the nitrogen-tetroxide tanks, and the oxidizer spilled out and was ignited by the flame from the rocket, which subsequently exploded. Rover no. 201 was one of a total of three units of this lunar vehicle.

Early Lunokhod test model during field trials.

The first test launch of the mighty N1 booster rocket took place on February 21, 1969. N1 no. 3L was the first to fly. Construction units 1L and 2L were used for ground tests and the training of handling crews. The payload for this mission was not a complete L3 unit, which would have consisted of the Block D upper stage, the LK lander, and the LOK orbital unit. All that was carried was Block D and a spacecraft with the designation 7K-L1S, a hybrid vehicle from the 7K-L1 and the 7K-LOK. A standard launch rescue system was used with it. It was envisioned that it would be a circumlunar mission without swinging into lunar orbit.

The limitation to this simplified payload was due in part to the fact that the LK lander was not yet ready for service and that they did not wish to risk a valuable complete L3 unit on this very first mission by the rocket. The main reason, however, was that the payload of the initial version of the N1 was not sufficient to put the 99-ton L3 lunar unit into Earth orbit.

Therefore, several of the measures for the later increase in performance were to be tested on this first flight, whose payload weight

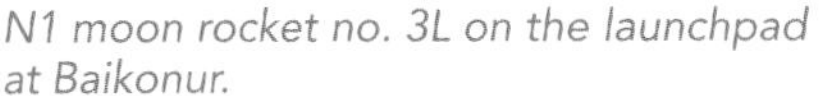

N1 moon rocket no. 3L on the launchpad at Baikonur.

Launch of N1 production unit 3L.

was restricted to 82.7 tons. Only the launch was included, however, and not the vehicle itself. The orbital inclination to the equator was therefore changed from the originally planned 65 degrees to 51.6 degrees. The orbital trajectory for the parking orbit around Earth was lowered from 180 to 132 miles.

Later flights were supposed to use fuels that had been deeply cooled, kerosene at –4 degrees F and liquid oxygen at –312 degrees F. The thrust of the thirty first-stage engines was to be increased by 2 percent, which was to be achieved by increasing combustion chamber pressure.

The mission began at 12:18:07 Moscow time and initially seemed to be successful, although two of the thirty engines shut down immediately after liftoff. However, the system was designed to withstand the loss of four rocket engines in the first stage. In the sixty-ninth second of flight, however, all the engines shut down. The rocket remained as a single unit and continued its flight path to apogee. Then it began the long fall back to Earth, until it finally crashed 32 miles from the launch site.

While there was much disappointment because of the fact that they had not had a test stand for the first stage, they were not completely dissatisfied with the result. In any event the rocket had operated for more than a minute, and the launchpad was undamaged. The launch rescue system had again functioned

Mission Data, First N1 Mission	
Mission designation	—
Date	February 21, 1969, 20:11 CET
Spacecraft	Soyuz 7K-L1S no. 1, rocket 3L
Booster rocket	N1
Spacecraft weight	approx. 165,345 pounds
Planned mission objective	Circumlunar flight. First test flight by the N1 booster rocket
Mission results	Multiple engine failures during initial launch phase. All engines shut down 69 seconds after liftoff
Fate	The rescue system brought the capsule to safety 21.75 miles from launch site

extremely well and delivered the crew cabin to safety. It was found 22 miles from the launch site, completely intact.

It was discovered that the crash was almost completely due to the shortcomings of the KORD engine control system (KORD = KOntrol Raketnykh Dvigateley, or digital engine control system). The engines themselves had functioned well, although at first they were high on the list of suspects during the investigation.

March 3: Apollo 9 completes a very successful ten-day mission in Earth orbit with astronauts McDivitt, Scott, and Schweickardt, during which the lunar lander is tested for the first time.

May 18: Apollo 10 undertakes the dress rehearsal for the moon landing. Astronauts Stafford, Young, and Cernan approach to within 48,000 feet of the lunar surface. The flight lasts eight days.

On June 14, 1969, the first attempted return mission to procure lunar soil samples failed, as did the second launch of an E8-5 spacecraft at the same time. The hope of bringing back lunar material to Earth before the Americans became increasingly unlikely. The fourth stage of the Proton booster rocket failed to ignite, and the probe burned up in Earth's atmosphere.

Mission Data, E-8-5 No. 402	
Mission designation	—
Date	June 14, 1969, 05:00 CET
Spacecraft	E-8-5 no. 402
Booster rocket	Proton K 8K78K
Spacecraft weight	12,345 pounds
Planned mission objective	Automatic sample return
Mission results	Third stage of the booster rocket failed to ignite
Fate	Total loss of rocket and payload

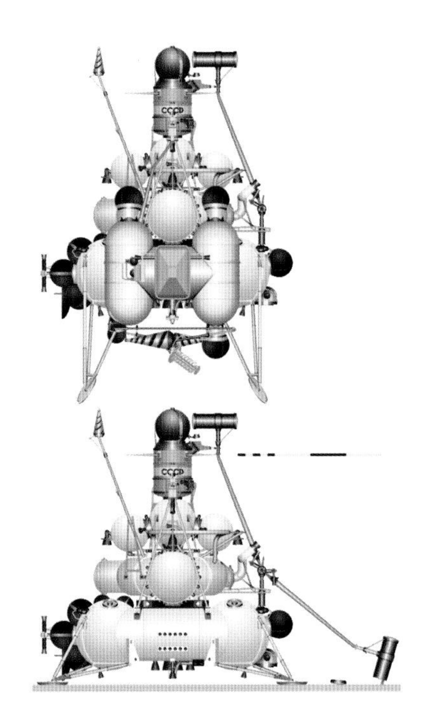

Diagram of Luna 16 in "folded" state as it was housed under the booster rocket's payload fairing, and with landing legs deployed.

1969: SECOND HALF OF THE YEAR

The two N1 launchpads at Baikonur were only about 1,500 feet apart.

After the crash of the first N1 booster rocket, the Soviets were relatively confident that they had identified and corrected the weaknesses of the system. Some modifications were introduced immediately; a larger number were deferred. In the meantime, the no. 4 production unit was taken from the operational cycle. It was to incorporate all the changes that had resulted from the flight of unit 3L and that would result from the flight of unit no. 5. It would thus be the first example of the modified N1 series.

Once again an L3S hybrid system was to be transported as payload instead of an L3 lunar lander complex. It was again based on an L1 Zond spacecraft with elements of the 7K-LOK spacecraft, functional Block G and Block D stages, and a dummy LK lander. Only the launch rescue system was completely up to standard. The mission was to be carried out as a circumlunar flight.

The ballisticians had calculated 23:18 Moscow time as the launch time necessary to precisely achieve the planned lunar trajectory. This never took place, however, since the second test flight of the N1 lasted just a few seconds. A quarter of a second before liftoff, with all thirty rocket engines running, engine no. 8 exploded and devastated the entire engine area. Despite this, the rocket lifted off. In a desperate attempt to get a grip on the unfortunate situation, in the seconds that followed, the KORD control unit shut down engines 7, 19, 20, and 21 one at a time. But the destruction of the rocket was unstoppable, as the damage, the torn lines and severed cables caused by the engine explosion, and the resulting fires spread.

The KORD system then shut down all the remaining engines except one, since, curiously, engine no. 18 continued running. After twelve seconds the rocket had reached a height of 650 feet above the launchpad and for an instant it stood still there. When engine no. 18, which was still running, slowly pushed the rocket into the horizontal, in the fifteenth second of the failed mission the launch rescue system activated itself and delivered the capsule to safety. Then the rocket, lying on its side, began to fall back to Earth and after twenty-three seconds crashed onto the launchpad.

The ensuing explosion created a bright flash that was seen in Leninsk (the present-day city of Baikonur), 21 miles away, and one and a half minutes later there was a deafening explosion. Doors and windows were blown

Mission Data, Second N1 Mission	
Mission designation	—
Date	July 3, 1969, 20:11 CET
Spacecraft	Soyuz 7K-L1S no. 2, rocket 5L
Booster rocket	N1
Spacecraft weight	approx. 165,345 pounds
Planned mission objective	Circumlunar flight. Second test flight by the N1 booster rocket
Mission results	Rocket exploded immediately after liftoff. The N1 launch installation 110 East was completely destroyed and was never rebuilt
Fate	The rescue system brought the capsule to safety

out of buildings up to 3.5 miles away. The launch complex was so completely leveled that it was never completely rebuilt. Miraculously, the second N1 launchpad just 1,500 feet away was relatively undamaged.

The second E-8-5 sample return probe was launched on July 13, 1969. It was the last attempt to bring lunar material to Earth before the Americans, and it just failed. In this case the failure was due less to technology than to the inadequate topographical material available to the Soviet engineers. A heavy price was paid for a series of planned orbital photo probes failing to reach their objectives.

Had the spacecraft carried out its mission as planned and had Apollo 11 failed, then the Soviet Union would have achieved a tremendous propaganda coup.

Four Proton K missions in a row had failed prior to the launch of the E-8-5. This time, however, all went well. Shortly after launch and achievement of lunar transfer trajectory, the spacecraft was given the designation Luna 15. The probe was on its way to the moon three days before Apollo 11, but the eyes of the world were not on it, but on the fate of Armstrong, Aldrin, and Collins.

On July 17 at 13:00 Moscow time, Luna 15 reached lunar orbit as planned. Apollo 11 was already en route to Earth's satellite. It took fifty-six orbits of the moon, however, for the decision makers in Moscow under Georgi Tyulin, deputy minister for machine building, to make the decision to initiate the landing. Prior to this there had been several delays because the data from the probe's radar altimeter were showing very rugged topography in the landing area.

When the Americans landed on the moon at 23:17 Moscow time on July 20, the mission managers of Luna 15 were still pondering their lunar maps. Finally, at 18:47 Moscow time, Luna 15 began its descent. At that time the American astronauts Armstrong and Aldrin were already preparing for the return to the lunar orbiter *Columbia*, which was to take place in two hours.

Luna 15's descent to the moon was supposed to take about six minutes and initially went as planned, but four minutes after the start of the landing ignition, to the controller's dismay, the signals from the probe suddenly went silent. The last telemetry data received from the probe showed an altitude of 9,000 feet and a speed of 300 miles per hour over the ground. It is now assumed that Luna 15 collided with a mountain that the Soviet scientists failed to take into consideration when selecting the landing trajectory.

The report by the TASS news agency was brief as usual: Luna 15's research program had been completed as planned after the space probe reached the anticipated area. There was one more small irony in the story: even if everything had gone exactly according to the predetermined mission plan for Luna 15, if the probe had collected soil samples, and if as planned it had begun the return flight,

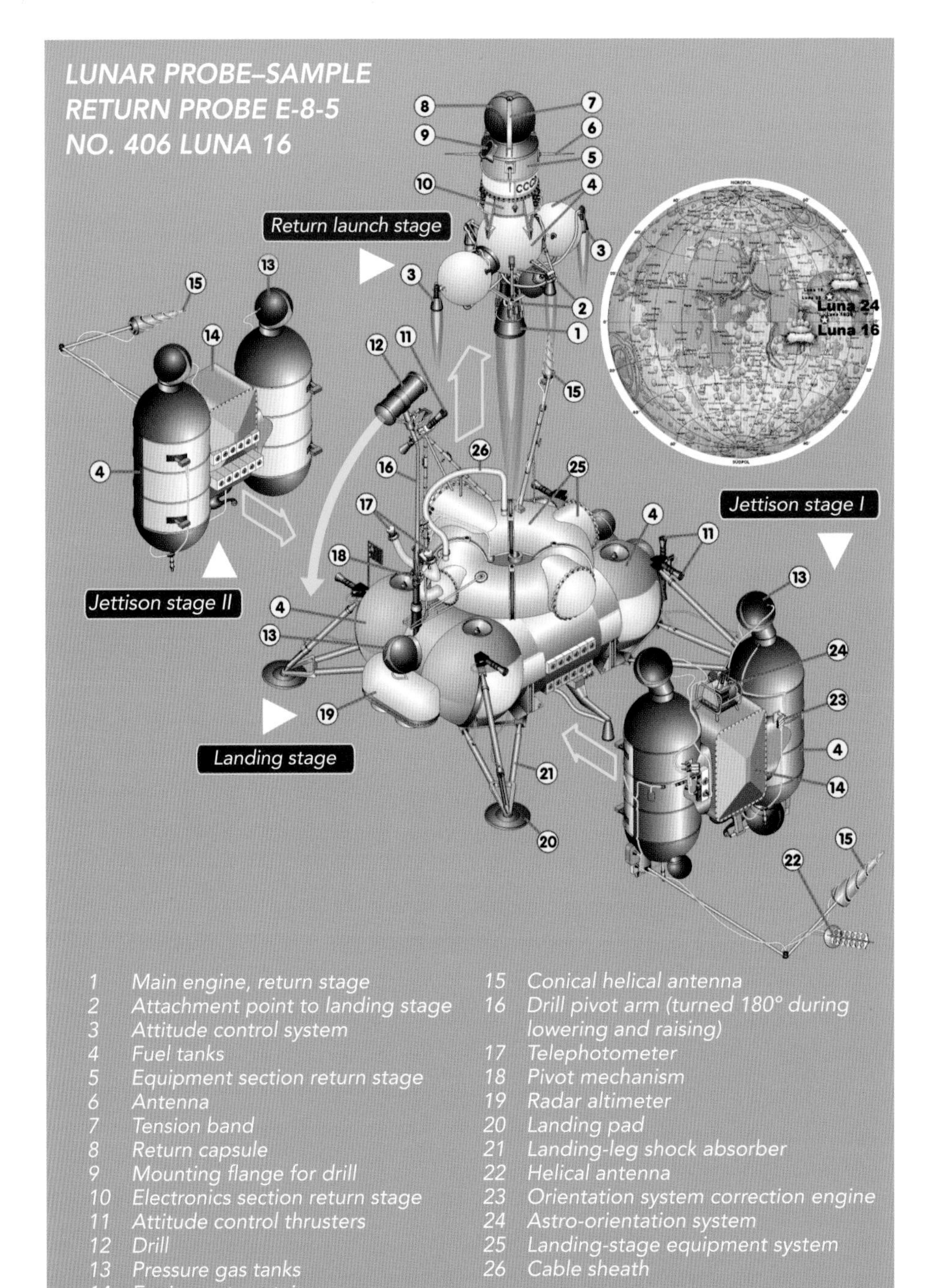
LUNAR PROBE–SAMPLE
RETURN PROBE E-8-5
NO. 406 LUNA 16
Return launch stage
Jettison stage II
Jettison stage I
Landing stage
Luna 24
Luna 16
1 Main engine, return stage
2 Attachment point to landing stage
3 Attitude control system
4 Fuel tanks
5 Equipment section return stage
6 Antenna
7 Tension band
8 Return capsule
9 Mounting flange for drill
10 Electronics section return stage
11 Attitude control thrusters
12 Drill
13 Pressure gas tanks
14 Equipment container
15 Conical helical antenna
16 Drill pivot arm (turned 180° during lowering and raising)
17 Telephotometer
18 Pivot mechanism
19 Radar altimeter
20 Landing pad
21 Landing-leg shock absorber
22 Helical antenna
23 Orientation system correction engine
24 Astro-orientation system
25 Landing-stage equipment system
26 Cable sheath

Luna 16 in final assembly.

even then the small capsule with the lunar material would have reached Soviet soil two hours and four minutes after the landing of Apollo 11. The race to the moon had been over before it had even begun.

Like all production probes of the E-8-5 series, Luna 15 consisted of two main components, the first of which was the landing stage, which carried out the course corrections on the way to the moon, braked the entire combination into lunar orbit, and finally carried out the landing. This stage had four landing legs and all the equipment necessary for the landing and subsequent ground operations.

The second main component was the ascent stage, a rocket independent of the landing stage that used the landing unit as a launchpad. Above on the ascent stage was the spherically shaped return capsule, and in it the hermetically sealed containers with the soil samples. The ascent stage made a direct flight to Earth without first orbiting the moon. The entire "landed" weight of the landing and ascent stages was a terrestrial 4,145 pounds. The total weight of the complete space probe, including fuel, when it separated from the Block D stage after the translunar burn maneuver was 12,960 pounds. This "initial weight" was roughly the same for all spacecraft of the E-8 series, no matter whether a sample return mission, a lunar orbit mission, or the depositing of a Lunokhod rover was carried out.

July 20: The first manned landing on the moon is a triumph for the Americans, who thus clearly win the space race. Astronauts Armstrong, Collins, and Aldrin fly to the moon in Apollo 11, with launch on July 16, and Armstrong and Aldrin land in the Sea of Tranquility four days later.

Mission Data, Luna 15	
Mission designation	Luna 15
Date	July 13, 1969, 03:55 CET
Spacecraft	E-8-5 no. 401
Booster rocket	Proton K 8K78K
Spacecraft weight	12,345 pounds
Planned mission objective	Automatic sample return
Mission results	Crashed in the Mare Crisium less than a minute before landing. Flew into a mountain during approach due to insufficient position information
Fate	Crashed at 17 degrees north latitude, 60 degrees east longitude

The circumlunar mission of LK-1 spacecraft no. 11, with the official designation of Zond 7, was completely successful. The mission began in the early morning hours of August 8, 1969. The probe reached its minimum distance from the moon on September 11. The landing on Earth, this time involving a successful "skip reentry," took place on August 14, 1969, south of the city of Qostanai in Kazakhstan. The pictures this probe brought back were very popular in the Soviet Union and were widely distributed. They were a little balm on the souls of the Soviet politicians and scientists who were forced to watch the grandiose success of the Americans, while their own efforts failed in almost every attempt.

On September 23, 1969, another sample return mission failed, as so often before because of the booster rocket. E-8-5 probe no. 403 reached Earth orbit, but the second ignition of the Block D stage, which was supposed to place it on the transfer trajectory, was unsuccessful. The Soviets named the spacecraft Cosmos 300. It burned up in Earth's atmosphere four days after launch.

On October 11, 1969, the spacecraft Soyuz 6 with Georgi Shonin and Valeri Kubasov

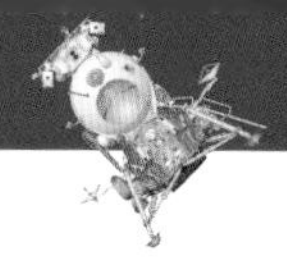

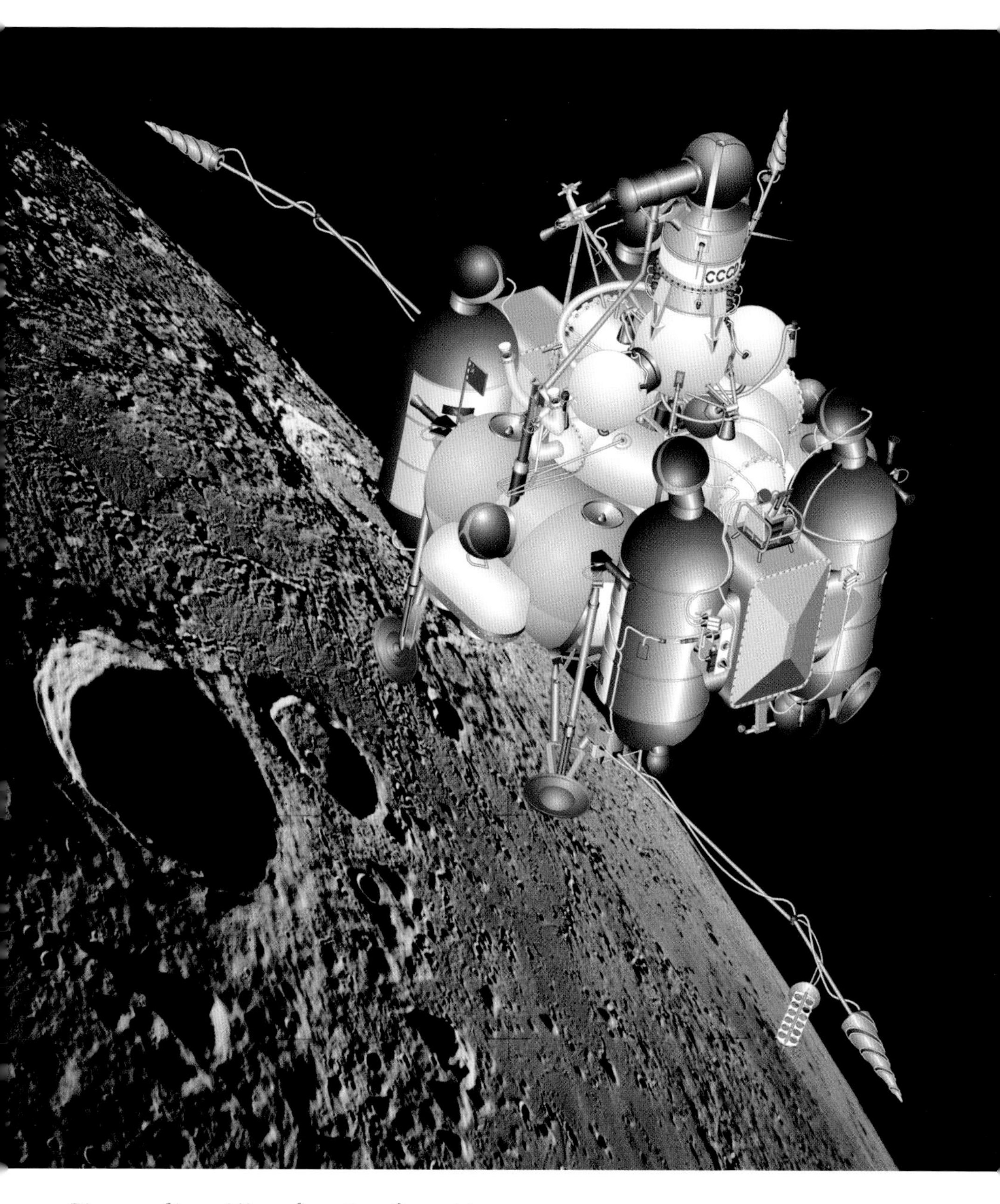

Diagram of Luna 16's configuration after arriving in lunar orbit. This illustration is of course also applicable to all subsequent sample return missions that made it to lunar orbit.

Mission Data, Zond 7	
Mission designation	Zond 7
Date	August 7, 1969
Spacecraft	Soyuz 7K-L1 no. 11
Booster rocket	Proton K 8K78K
Spacecraft weight	13,183 pounds
Planned mission objective	Circumlunar flight. Test of Zond spacecraft
Mission results	Mission successful. Flew around the moon at a distance of 1,233 miles. First fully successful flight that even a crew would have survived undamaged
Current location	The Zond 7 capsule can be seen in the museum in Orevo

Mission Data, Cosmos 300	
Mission designation	Cosmos 300
Date	September 23, 1969
Spacecraft	E-8-5 no. 403
Booster rocket	Proton K 8K78K
Spacecraft weight	12,345 pounds
Planned mission objective	Automatic sample return
Mission results	The spacecraft became stranded in low Earth orbit because the engine of the Block D upper stage did not ignite for the TLI maneuver
Fate	Crashed after a few days

onboard was launched. Just one day later followed Soyuz 7 with Anatoli Filipchenko, Vladislav Volkhov, and Viktor Garbatko, and one day after that Soyuz 8 with Vladimir Shatalov and Alexey Yeliseyev. The objective of the mission was the docking of two spacecraft while it was filmed from the third Soyuz.

The docking maneuver between Soyuz 7 and Soyuz 8 was supposed to take place on October 14, with Soyuz 7 as the active

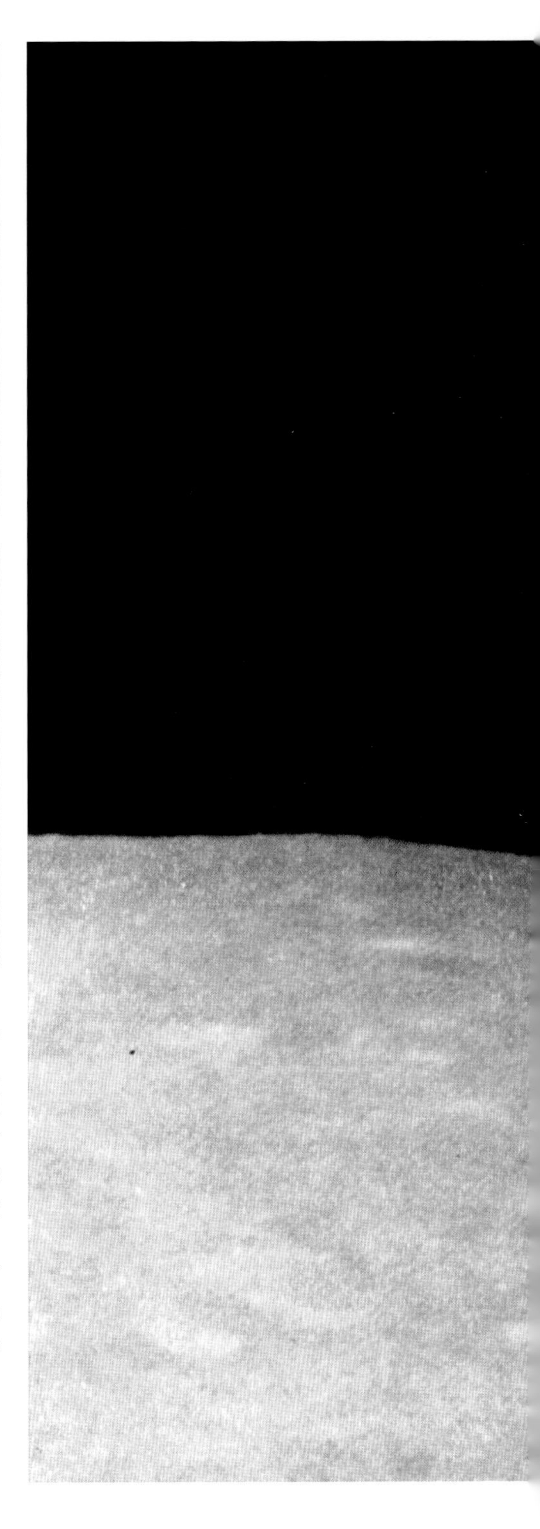

Earth over the moon, photographed from Zond 7 from a distance of about 1,200 miles above the surface of the moon.

spacecraft. The rendezvous began at a distance of 150 miles, and the two spacecraft approached to within about 2,300 feet. The crews could already see each other, but then the automatic rendezvous system failed. Shatalov then requested permission for a manual rendezvous maneuver, but while the discussion between the controllers on the ground and the crew in orbit went back and forth, the two spacecraft drifted apart. Failure of the Igla system also meant that the crew received no track information for a manual maneuver from the system, making it practically impossible to complete the rendezvous maneuver by hand. On October 15, after a series of maneuvers, the two spacecraft again came within about 10 miles of each other.

Despite the fact that for the first time there were three spacecraft and their space travelers in space at the same time, the mission had to be considered a failure. The three spacecraft landed one per day, after each had spent five days in space.

During this triple mission, hardware and procedures necessary for a manned Soviet moon landing were tested for the last time. Even after two years of great effort, Soviet engineers had not been able to equip the Soyuz with instrumentation that would have allowed the cosmonauts to carry out a successful manual-docking maneuver. The failure of this mission was one of the last nails in the coffin of the Soviet manned lunar-landing program.

On October 22, 1969, another sample-collecting mission ended in failure because of the booster rocket, and once again it was Block D that failed, this time during the first ignition. As a result, the spacecraft, designated Cosmos 305, fell back to Earth within one orbit.

Mission Data, Cosmos 305	
Mission designation	Cosmos 305
Date	October 22, 1969
Spacecraft	E-8-5 no. 404
Booster rocket	Proton K 8K78K
Spacecraft weight	12,345 pounds
Planned mission objective	Automatic sample return
Mission results	Exactly the same fault as four weeks earlier. Loss of mission
Fate	Crashed after just one orbit of Earth

November 14: The second American lunar expedition with astronauts Conrad, Gordon, and Bean sets off on a ten-day trip to the Sea of Storms. The mission is extremely successful.

On November 28, 1969, the Proton K failed again. The test flight of the version of the Block D upper stage selected for the N1 rocket failed because of a broken fuel line caused by excessive vibration in the 556th second of flight.

1970

On February 6, 1970, the Soviet Union again attempted to launch a soil sample collection mission to the moon. A pressure sensor in the first stage of the Proton booster rocket shut it down too soon, and the probe failed to even reach a parking orbit.

The Proton K booster rocket was the clear weak spot, both in the L2 program for launching E-8 probes and in the L1 program for launching

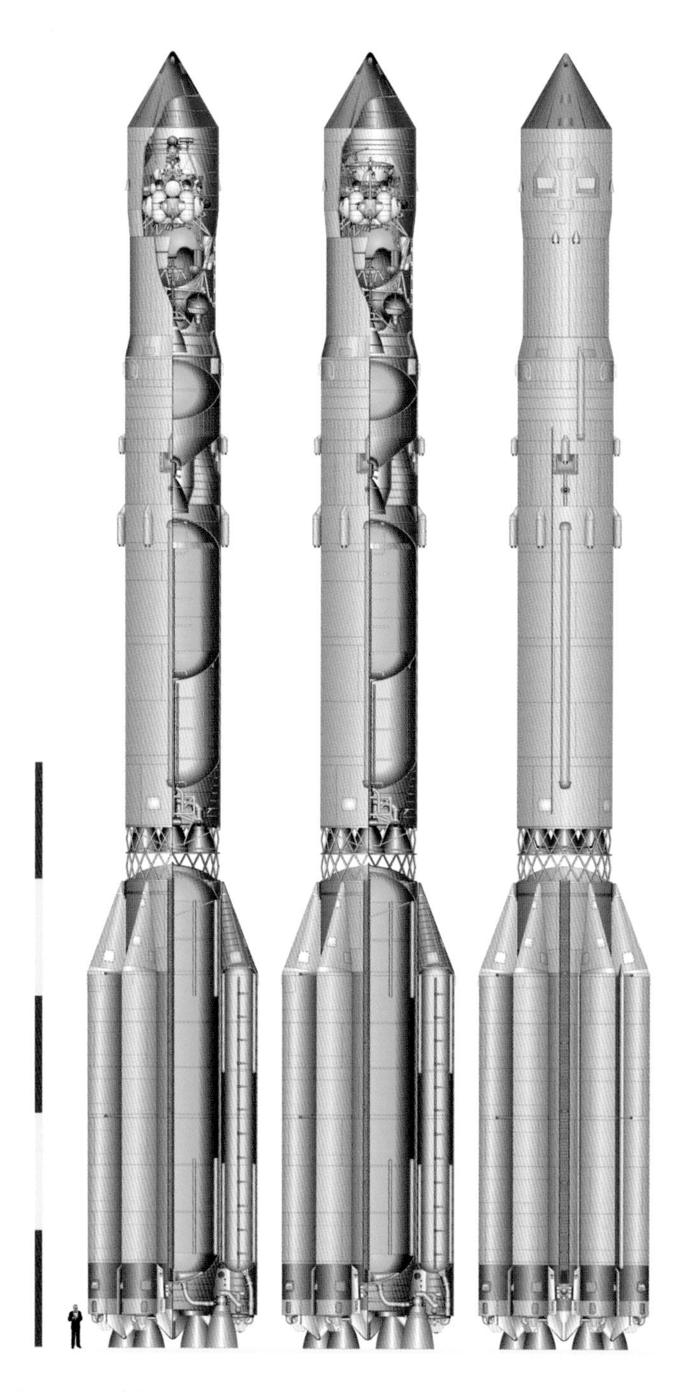

Diagram of the Proton K for the E-8 series probes. Left, landing probe; center, Lunokhod.

Mission Data, E-8-5 No. 405	
Mission designation	—
Date	February 6, 1970
Spacecraft	E-8-5 no. 405
Booster rocket	Proton K 8K78K
Spacecraft weight	12,345 pounds
Planned mission objective	Automatic sample return
Mission results	The first-stage engine of the Proton was shut down after 128 seconds because of a false sensor reading. Loss of the mission
Fate	Crashed about 120 miles from the launch site

the Zond spacecraft. Even by the standards of the time, it had the worst statistics for all orbital boosters developed by any spacefaring nation. Between March 10, 1967, and February 6, 1970, the four-stage variant of the Proton was launched nineteen times and failed an unbelievable thirteen times. Only six times did it deliver its payload on the anticipated trajectory. Rocket after rocket failed, and the company that made them, Vladimir Chelomei's ZKBM (the former OKB-52), was pressed to improve the reliability of its booster rocket. All flights by the Proton were suspended between February and August 1970, and during that time many systems were redesigned, and the changes were tested in a flight made on August 18. Even afterward the rocket still had issues, but the failure rate dropped significantly.

April 11: The third American moon-landing mission fails, when halfway to the moon an oxygen tank explodes on Apollo 13's service module. In a dramatic rescue effort, NASA manages to bring astronauts James Lovell, Jack Swigert, and Fred Haise back to Earth.

On June 19, 1970, Soyuz 9 was launched with cosmonauts Nikolayev and Sevastyanov on an eighteen-day mission. With this long-duration flight the manned Soviet space program finally turned away from the moon and shifted its efforts to a new objective, the operation of a manned space station.

Between September 1970 and June 1971, at a time when the entire lunar program was being questioned more and more politically, successes began to arrive. It was noticeable that now, after losing the race to the moon to the Americans, the enormous pressure and terrible haste began to subside. Something began that in sport one calls "a roll." Whereas before there had been failure after failure, now there was success after success. The "run" did not end until the tragic events at the end of the Soyuz 11 mission and the last flight tests of the N1 super rocket, with which the Soviet Union had once intended to beat the Americans to the moon.

The "run" began on September 12, 1970, at 14:25 Central European Time. That day a Proton K delivered the fifth E-8-5 space probe to the moon without problems. Soon after launch the spacecraft was given the name Luna 16, and it was to achieve the success that had eluded Luna 15 a year earlier. On September 13, there was a midcourse correction and at 00:38 on December 17, the probe entered an almost circular lunar orbit at an altitude of 66 miles and an inclination of 70 degrees to the lunar equator.

On September 19, Luna 16 changed its orbital altitude and inclination. From then on it traveled in an elliptical orbit with a perigee of 9 miles, an apogee of 66 miles, and an inclination of 71 degrees.

At 06:12 on September 20, the probe began its descent to the surface of the moon with the ignition of the engine. About six

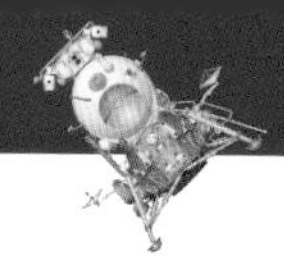

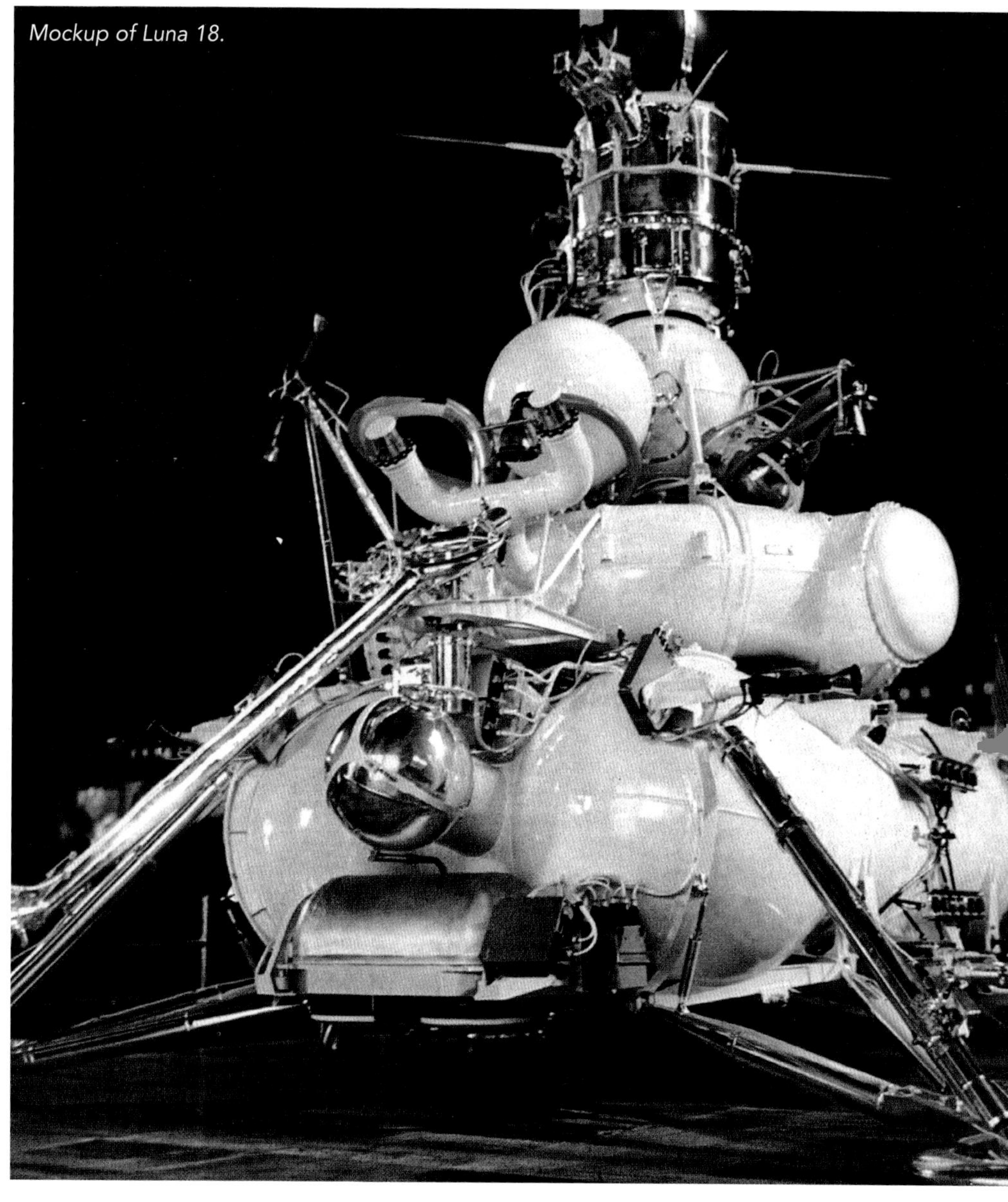
Mockup of Luna 18.

minutes later it was sitting in the Mare Fecundidatis, the Sea of Fertility, at a position of 0.68 degrees south altitude and 53.6 degrees east longitude. Seventy-two minutes later, Luna 16 began removing a core sample from the landing site and then stowed it in the sample container onboard the return section of the probe. The sample was 14 inches long and weighed 3.55 ounces. The entire process took no longer than seven minutes.

On September 21, at 08:43 Central European Time, Luna 16's ascent engine fired and

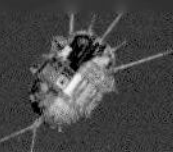

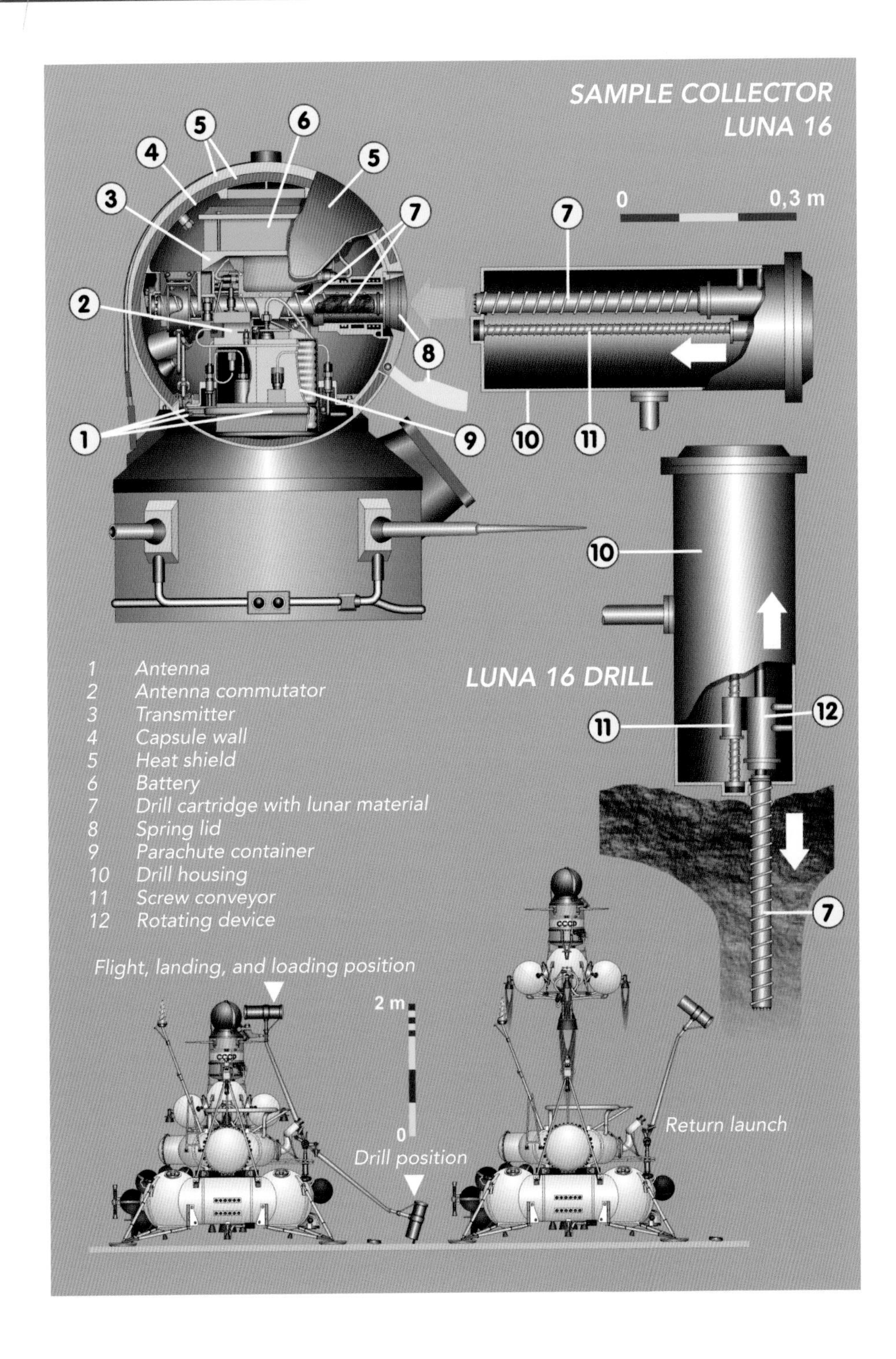
SAMPLE COLLECTOR
LUNA 16
0
0,3 m
1
2
3
4
5
6
7
8
9
10
11
12
LUNA 16 DRILL
1 Antenna
2 Antenna commutator
3 Transmitter
4 Capsule wall
5 Heat shield
6 Battery
7 Drill cartridge with lunar material
8 Spring lid
9 Parachute container
10 Drill housing
11 Screw conveyor
12 Rotating device
Flight, landing, and loading position
2 m
0
Drill position
Return launch

the probe began its direct return to Earth. At 06:10 on September 24, the return capsule separated from the transfer module. Distance to Earth was 29,825 miles.

At 06:26 Central European Time, the probe landed in central Kazakhstan 50 miles east of the city of Shesqasghan. For the first time in the world, an unmanned space vehicle had brought lunar material back to Earth. Altogether, however, it was the third return of samples, since Apollo 11 and 12 had previously brought back about 130 pounds of moon rocks. This success was balm on the wounds of the Soviet space engineers, who looked back on along series of failures.

These plaques were on Luna 16.

The circumlunar mission of Zond 8 began on October 20, 1970, and was a complete success. On October 21, the probe sent back pictures of Earth from a distance of 40,000 miles. On October 24, Zond 8 flew around the moon at a distance of about 600 miles. It landed in the Indian Ocean on October 27, 15 miles from the recovery vessel *Taman*.

With its lunch on November 10, 1970, began the success story of Luna 17, without a doubt one of the high points of the Soviet space program along with the sample return by Luna 16. On November 16, the space probe swung into orbit around the moon, and two days later it landed in the Mare Imbrium. Its payload this time was not an ascent stage with an Earth return capsule; instead it was the 1,667-pound Lunokhod 1 rover. Half an hour after the landing, the control center established first contact with Lunokhod. Another fifty minutes later the vehicle sent its first pictures of the lunar surface. At that time it was still on the back of the Luna 17 landing probe.

Mission Data, Luna 16	
Mission designation	Luna 16
Date	September 12, 1970
Spacecraft	E-8-5 no. 406
Booster rocket	Proton K 8K78K
Spacecraft weight	12,345 pounds, lander 4,145 pounds
Planned mission objective	Automatic sample return
Mission results	Landing on September 20, 1970. Return launch to Earth with 3.6 ounces of lunar material on September 21. Landing on Earth on September 24, 1970
Current location of landing probe	0 degrees, 41 seconds south latitude; 56 degrees, 18 seconds east longitude in the northeastern Mare Fecunditatis

Luna 16's return capsule after landing in Kazakhstan.

Three hours after landing, orders radioed from Earth initiated the unfolding of two ramps, over which Lunokhod 1 drove down to the surface of the moon. The rover remained active until September 14, 1971, covering 6.5 miles. It sent back 20,000 photographs and 200 panoramic images and undertook several hundred soil examinations.

Mission Data, Zond 8	
Mission designation	Zond 8
Date	October 27, 1970
Spacecraft	Soyuz 7K-L1 no. 14
Booster rocket	Proton K 8K78K
Spacecraft weight	11,850 pounds
Planned mission objective	Circumlunar flight. Test of Zond spacecraft
Mission results	Mission successful. Flew around the moon at a distance of 690 miles. Landed in Indian Ocean on October 27, 1970
Last contact	—
Current location	Not known

The total "landed" weight of Luna 8, including Lunokhod 1, was 4,050 pounds. The rather bathtub-like appearance of Lunokhod 1 was due to the fact that most of the instrumentation and operating systems were pressurized. The pressure gas was nitrogen, and the internal pressure was one atmosphere. Over the main body was a sort of cover plate. It was open during the lunar day, which was fourteen Earth days long. On the inside were the solar cells, with which the batteries could be charged. During the lunar night the cover was closed and provided added insulation. Another source of heat for the lunar night,

and also an additional source of energy, was a radioisotope generator powered by Polonium 210, weighing 22 pounds.

At the end of a lunar day, Lunokhod was parked so that at the start of a new day the solar cells received maximum energy. Under the best conditions they produced about a kilowatt of electric power.

Each of the eight wheels had its own electric motor, so that the loss of one wheel did not endanger the mission. The rover was capable of dealing with the loss of two wheels, one on each side. Lunokhod was controlled by a team of five drivers sitting at monitor screens in Moscow. They could move the vehicle over the moon at a notable 650 feet per hour.

One special feature of the E-8 series was its two jettisonable reserve fuel tanks, which after the swing into lunar orbit were left there. This reduced weight for landing. The landing engine of the Luna probe was an inspired design by Isayev's OKB-2. The 11D417 rocket engine's thrust could be regulated between 1,650 and 4,270 pounds.

In the history of the Soviet lunar program, pervaded by so many failures, another beacon stands out that is deserving of special mention: the LK moon lander of the L3 complex. Although it was a highly complex machine, its development took place in the shadows of the other system's components, which were plagued by difficulties, without significant difficulties and very inconspicuously. At a time

Lunokhod 1 photographed during its ascent stage.

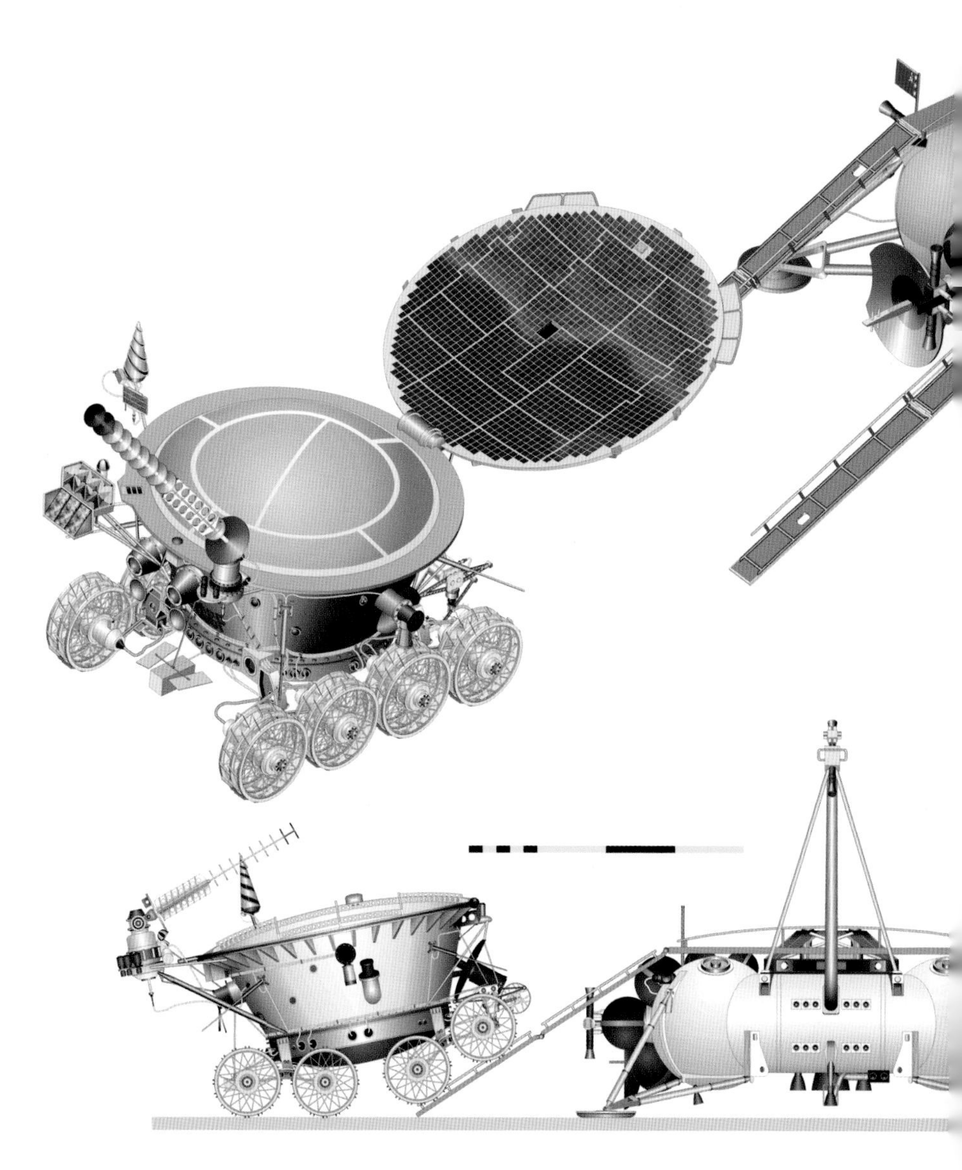

Lunokhod 1.

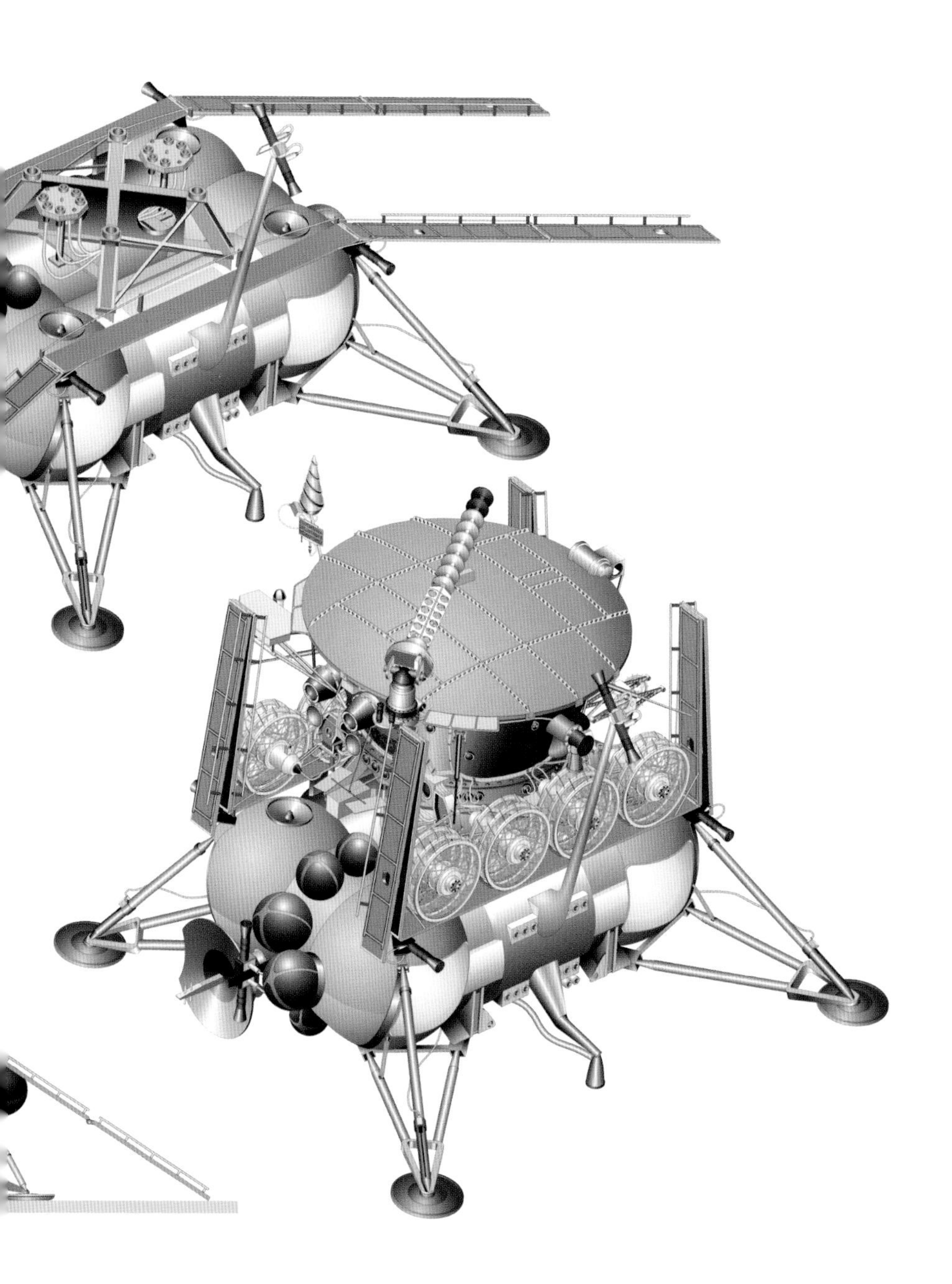

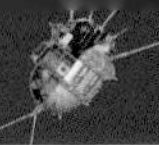

Landing site of Luna 17 with tracks left by Lunokhod 1.

Mockup of Lunokhod 1.

when the other systems were still pushing mountains of problems before them, it was the only really fully developed component, which also functioned well.

A special version of the Soyuz booster rocket with the designation 11A511L was procured for flight testing of the LK lander. It stood out mainly on account of its uniquely shaped payload fairing, beneath which the LK lander was housed for its flight trials in Earth orbit. Only three of this special version were built.

The first flight test of the LK lander began on November 24, 1970, under the designation Cosmos 379. Because a landing on the moon was not necessary for the tests, the lander

Mission Data, Lunokhod 1	
Mission designation	Luna 17
Date	November 10, 1970
Spacecraft	E-8EL no. 203 Lunokhod
Booster rocket	Proton K 8K78K
Spacecraft weight	12,566 pounds, rover 1,667 pounds
Planned mission objective	Mobile moon exploration by Lunokhod rover
Mission results	Landing on November 17, 1970; last contact on September 14, 1971; distance covered 6.55 miles
Current location	38 degrees, 17 seconds north latitude; 35 degrees, west longitude in the Mare Imbrium

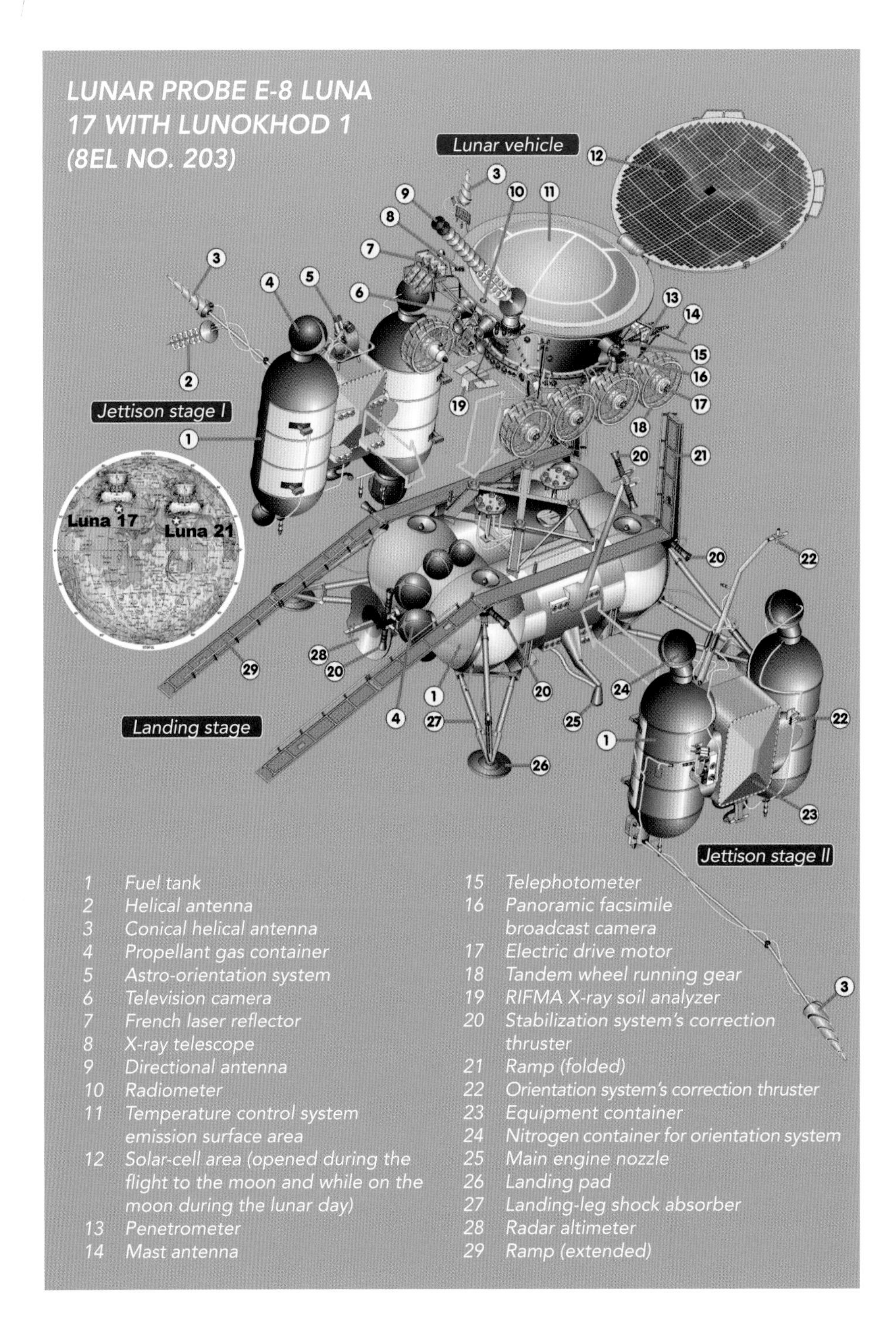
LUNAR PROBE E-8 LUNA 17 WITH LUNOKHOD 1 (8EL NO. 203)
Lunar vehicle
Jettison stage I
Luna 17
Luna 21
Landing stage
Jettison stage II
1 Fuel tank
2 Helical antenna
3 Conical helical antenna
4 Propellant gas container
5 Astro-orientation system
6 Television camera
7 French laser reflector
8 X-ray telescope
9 Directional antenna
10 Radiometer
11 Temperature control system emission surface area
12 Solar-cell area (opened during the flight to the moon and while on the moon during the lunar day)
13 Penetrometer
14 Mast antenna
15 Telephotometer
16 Panoramic facsimile broadcast camera
17 Electric drive motor
18 Tandem wheel running gear
19 RIFMA X-ray soil analyzer
20 Stabilization system's correction thruster
21 Ramp (folded)
22 Orientation system's correction thruster
23 Equipment container
24 Nitrogen container for orientation system
25 Main engine nozzle
26 Landing pad
27 Landing-leg shock absorber
28 Radar altimeter
29 Ramp (extended)

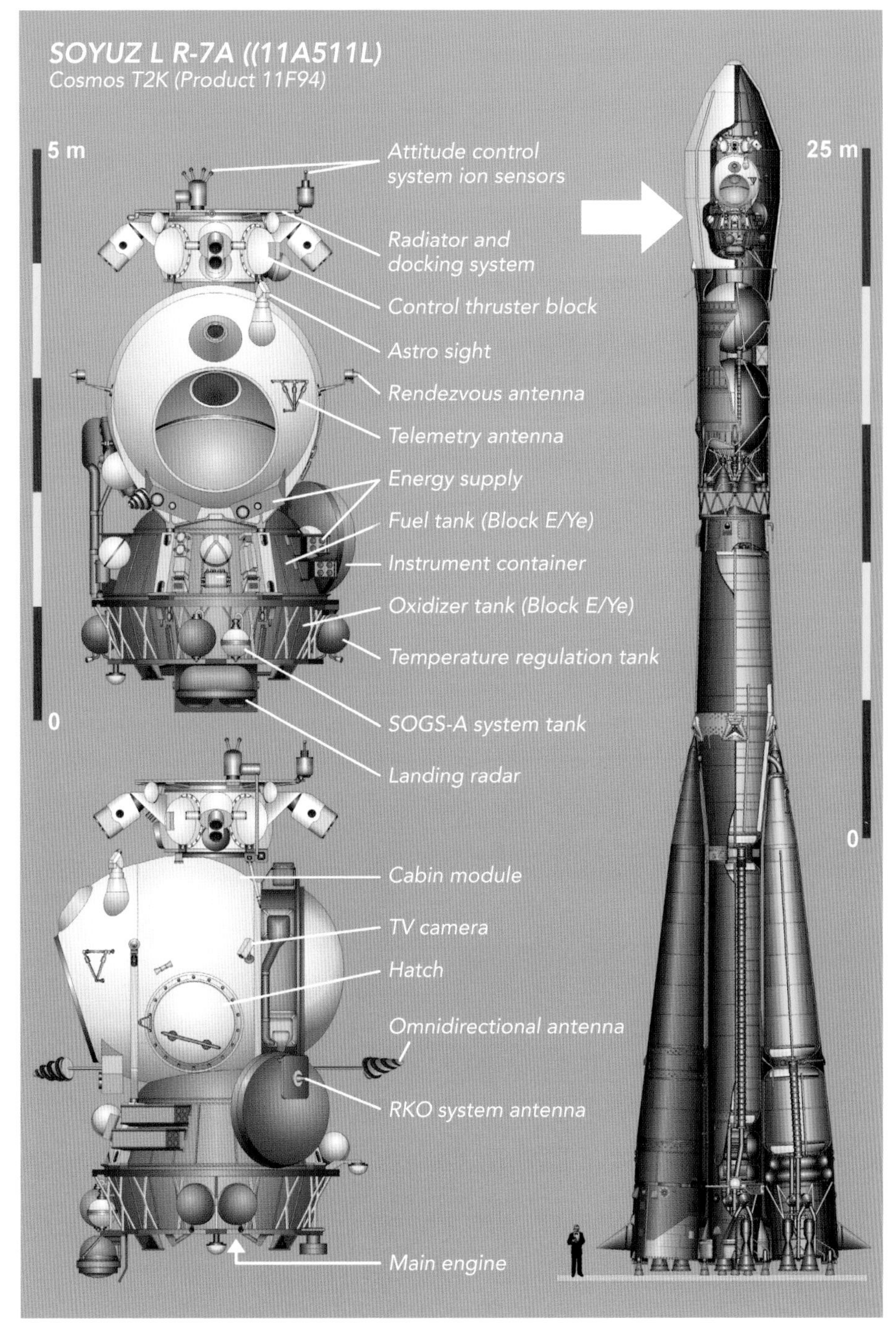
SOYUZ L R-7A ((11A511L)
Cosmos T2K (Product 11F94)
5 m
0
25 m
0
Attitude control system ion sensors
Radiator and docking system
Control thruster block
Astro sight
Rendezvous antenna
Telemetry antenna
Energy supply
Fuel tank (Block E/Ye)
Instrument container
Oxidizer tank (Block E/Ye)
Temperature regulation tank
SOGS-A system tank
Landing radar
Cabin module
TV camera
Hatch
Omnidirectional antenna
RKO system antenna
Main engine

had no landing legs. This corresponded to the mission of Apollo 5 by the Americans on January 22, 1968, in which the US lunar lander was tested unmanned in orbit. The MM lander also had no landing gear.

The mission of Cosmos 379 tested a complete moon flight. After insertion into Earth orbit on November 24, 1970, initially a three-day transfer phase to the moon was simulated. This was followed by hovering flight and a landing maneuver. During this, the apogee increased to 751 miles. Then a period on the moon was simulated. The LK lander's maximum duration on the moon's surface would have been twelve to sixteen hours. Afterward, Cosmos 379 ignited its ascent engine and increased speed by 0.93 miles per second. This was exactly the velocity that would have been needed to return from the surface of the moon to lunar orbit. Then followed a rendezvous and docking maneuver with many brief engine ignitions. The entire mission was a perfect success.

1971

January 31: The crew of Apollo 14 (Shepard, Roosa, Mitchell) fly to the moon. The very successful mission's landing is in the Frau Mauro Highland. The mission lasts nine days.

The second unmanned test flight of the LK moon lander in Earth orbit began on February 26, 1971. The Soviet Union designated the spacecraft Cosmos 398. The flight maneuvers were very similar to those of Cosmos 379, and this mission too was almost trouble free.

But this was the end of the good news that had prevailed since September 1970. June 1971 was to be one of the blackest months in the history of Soviet spaceflight. It began with a pleasing event, when on June 6 the manned spacecraft Soyuz 11 was launched successfully. Only hours later it docked with the space station Salyut 11, and the crew transferred to the space station. The Soviet Union had thus achieved another first, not in the moon-landing program but at least in Earth orbit. It was the first time in the history of humanity that a manned space station had been placed in operation. Three weeks later the mission of Georgi Dobrovolski, Viktor Patsayev, and Vladislav Volkhov was to end tragically.

On June 25, Alexey Isayev, with Sergey Korolev one of the most important personalities in the Soviet space program, died unexpectedly. Isayev was head of OKB-2. He was responsible for almost all the engines of the manned Soviet spacecraft and the engines of the Mars, Venera, and Luna space probes. His funeral took place on June 28, 1971, at the Novodevichi cemetery of honor in Moscow.

Just hours earlier, on June 27, the third experimental flight of an N1 rocket also failed. Having become cautious after the failures if the two previous launches, this time the rocket carried dummy payloads: a Soyuz 7K-LOK mockup and a model of the LK lander. Since the main purpose of this flight was to test the first three stages, there was no need to keep to a lunar launch window. The mission therefore began at 02:15 Moscow time. At the launch site in Kazakhstan it was two hours later, and at that time of year it was already light.

The booster ran into difficulties as soon as the launch began. It was immediately apparent that there was some problem with the roll

This grainy image is an absolute rarity. It shows the launch of a Soyuz special version (11A511L) for flight trials of the LK lander in Earth orbit. Also see the illustration on page 141.

control system, since the N1 began, slowly at first then ever more quickly, to rotate about its longitudinal axis. The correction thrusters were unable to compensate for this movement. Forty-eight seconds after liftoff this rolling movement was so great that pieces began to separate from the rocket. Two seconds later the KORD system shut down all the engines.

The spinning rocket then began to break up. The largest pieces of wreckage fell 15 miles from the launchpad and created a large crater with a diameter of 100 feet and a depth of 50 feet. The rest of the rocket was scattered over the entire flight path.

Three days after this latest failed launch by the N1, the worst disaster in the manned

Mission Data, Third N1 Mission	
Mission designation	—
Date	June 27, 1971
Spacecraft	Soyuz 7K-LOK
Booster rocket	N1
Spacecraft weight	approx. 165,346 pounds
Planned mission objective	Third test flight by an N1 booster rocket
Mission results	Rocket exploded about fifty seconds after liftoff
Fate	The rescue system brought the capsule to safety

space program of the USSR occurred. For three weeks, television viewers in the Soviet Union and the world had witnessed cosmonauts Dobrovolsky, Patsayev, and Volkhov making almost daily reports from the space station Salyut 1. The three cosmonauts were on everyone's lips and were the heroes of the nation.

When they left the space station, they had set a new record for the longest stay in space, almost twenty-four days. Their return to Earth on June 30 seemed to go according to plan, even though there were several problems during separation from the station. Ground control was in contact with the crew until about 03:45. Then the retros fired and shortly afterward the landing capsule separated from the service module and orbital module. At first, ground control was not certain whether this event had actually happened, because since the time of retro ignition there had been no contact with the cosmonauts. Radar confirmed, however, that separation had taken place according to plan, and a regular reentry followed. In the control center, therefore, there was no great concern about the absence of radio contact, with only a minor problem with the communications system being suspected. This impression strengthened when the capsule was seen descending beneath its parachute. Soyuz 11 landed in the designated target area at 04:18.

However, when the recovery team opened the capsule's hatch, they were gripped by absolute horror. The three cosmonauts lay dead in their seats, their faces blotched dark blue. They were not strapped into their seats, and Dobrovolsky was entangled in his straps. Attempts to resuscitate them began immediately but were unsuccessful. The spacecraft had developed a leak at the start of the descent phase. The crew had obviously tried in their final seconds to find where their capsule was losing pressure. As was usual at that time (and with the then-current versions of the Soyuz capsules, being the only possibility with a crew of three), the crew were not wearing spacesuits.

July 26: Apollo 15, with astronauts Scott, Worden, and Irwin, launched to the moon. Scott and Irwin land near the Hadley Rille and spend almost three days on the moon.

On August 12, 1971, the third and last test of the LK landing vehicle went perfectly. The Soviet news agency TASS designated the spacecraft Cosmos 434. The test parameters for this final flight were once again tightened. This mission would have resulted in the Soviet lunar lander being released for the manned mission to the moon, but by this time Moscow's lunar-landing program was already in its last legs.

On September 2, 1971, at 14:41 Central European Time, the Soviets launched Luna 18 (E-8-5 no. 407) on another sample return mission. The descent to the surface of the moon was initiated on September 11, during the probe's fifty-fourth orbit of the moon. All seemed to go according to plan, but transmissions stopped at the moment the probe touched down, at 08:48 Central European Time. It was suspected that the probe had struck a rock during landing. The mission was the seventh attempt (counting the probes of the E-6 series) to retrieve soil samples from the moon.

On September 28, 1971, the Soviet Union launched the lunar orbital probe Luna 19. The spacecraft was the first of two advanced lunar orbiters based on the E-8 platform. It was thus

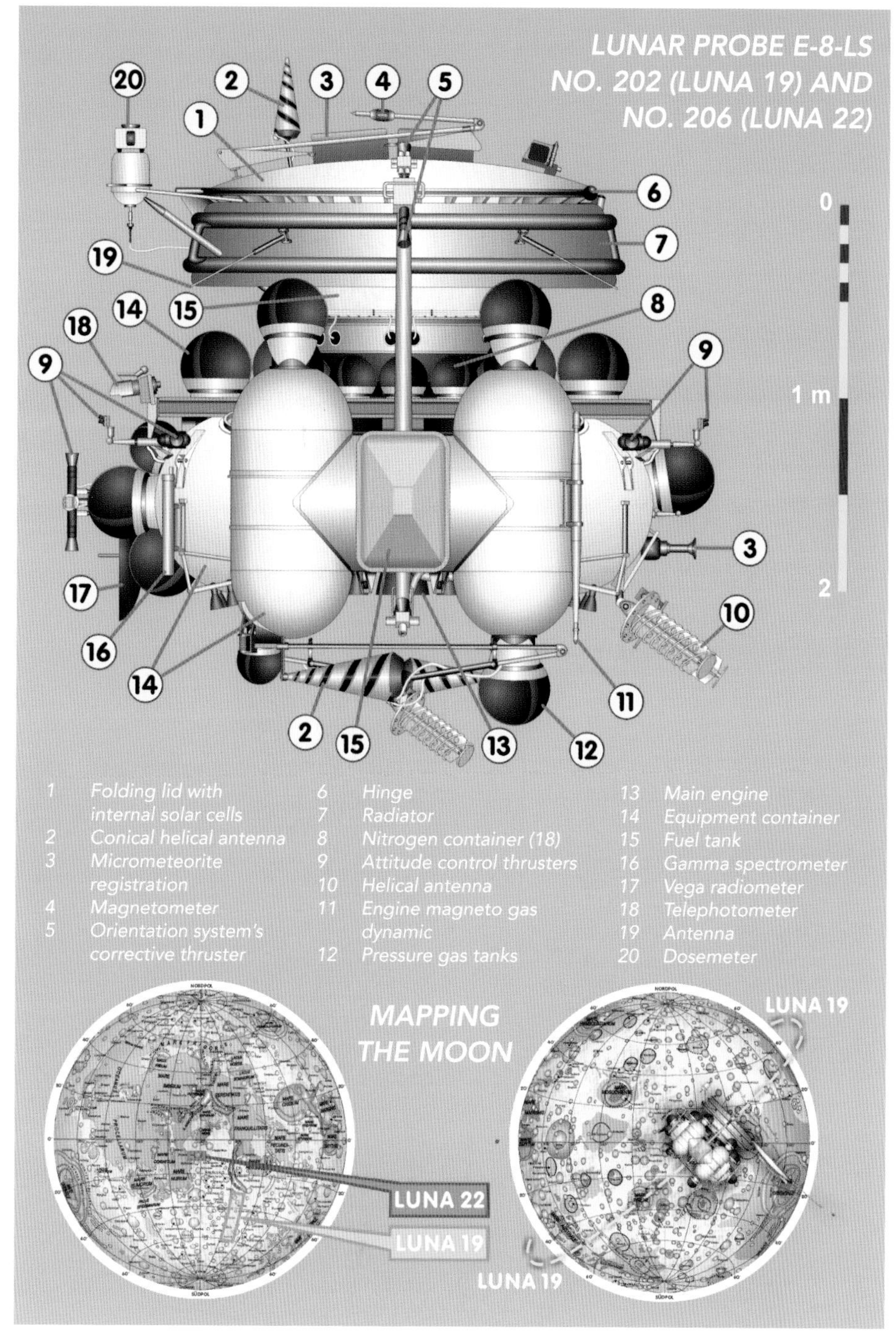

LUNAR PROBE E-8-LS
NO. 202 (LUNA 19) AND
NO. 206 (LUNA 22)
0
1 m
2
1 Folding lid with internal solar cells
2 Conical helical antenna
3 Micrometeorite registration
4 Magnetometer
5 Orientation system's corrective thruster
6 Hinge
7 Radiator
8 Nitrogen container (18)
9 Attitude control thrusters
10 Helical antenna
11 Engine magneto gas dynamic
12 Pressure gas tanks
13 Main engine
14 Equipment container
15 Fuel tank
16 Gamma spectrometer
17 Vega radiometer
18 Telephotometer
19 Antenna
20 Dosemeter
MAPPING
THE MOON
LUNA 22
LUNA 19
LUNA 19
LUNA 19

Mission Data, Luna 18	
Mission designation	Luna 18
Date	September 2, 1971, 14:40 CET
Spacecraft	E-8-5 no. 407
Booster rocket	Proton K 8K78K
Spacecraft weight	12,676 pounds, lander 4,145 pounds
Planned mission objective	Automatic sample return
Mission results	Contact lost at time of touchdown, at 08:48 on September 11. Probably caused by contact with a rock on the ground
Fate	3 degrees, 34 seconds north latitude; 56 degrees, 30 seconds east longitude at the edge of the Mare Fecunditatis

Mission Data, Luna 19	
Mission designation	Luna 19
Date	September 28, 1971, 11:00 CET
Spacecraft	E-8-LS no. 202
Booster rocket	Proton K 8K78K
Spacecraft weight	12,566 pounds
Planned mission objective	Lunar orbiter
Mission results	Insertion into lunar orbit on October 3, 1971. Orbit 87 x 87 miles, inclination 40.58 degrees
Last contact	Was deactivated on October 20, 1972. Crashed onto the surface of the moon. Date not known
Current location	Not known

the third type of space probe developed on that basis. Elements of the Lunokhod rover were also used. Almost the entire scientific payload was in the interior of a nitrogen-filled pressure vessel that had been developed for Lunokhod as an instrument and subsystem container. Luna 19 carried out a very successful mission, which did not end until October 1972.

1972

On February 14, the Soviet Union launched the sample return probe Luna 20 to the moon. It was planned that this probe would complete the mission that Luna 18 had failed to carry out. On February 18, it reached a circular lunar orbit with an orbital altitude of 62 miles and an inclination of 65 degrees to the lunar equator. At 20:13 Central European Time, Luna 20 fired its descent engine for four minutes and twenty-seven seconds and landed in the Sea of Fertility, just 75 miles from where Luna 16 had landed, and just 1.1 miles from where Luna 18 had crashed. At 23:58 Central European Time on February 22, the probe took off from the surface of the moon with one ounce of lunar material onboard. The return capsule landed safely in the Soviet Union on February 25.

April 16: The US launches its penultimate manned mission to the moon. The crew of Apollo 16 was made up of astronauts Young, Mattingly, and Duke. Young and Duke landed in the Descartes Highland and remained there for almost three days. On April 27, the crew returned safely to Earth.

The landing capsule of Luna 20

Mission Data, Luna 20	
Mission designation	Luna 20
Date	February 14, 1971, 04:28 CET
Spacecraft	E-8-5 no. 408
Booster rocket	Proton K 8K78K
Spacecraft weight	12,625 pounds, lander 4,145 pounds
Planned mission objective	Automatic sample return
Mission results	Landed on February 21, 1972. Return launch to Earth with 1.05 ounces of lunar material on February 22. Landing on Earth on February 25
Fate	Landed just 1.2 miles from the spot where Luna 18 landed

Mission Data, Last Flight of the N1	
Mission designation	N1/7L
Date	November 23, 1971, 09:12 CET
Spacecraft	Soyuz 7K-LOK 6A
Booster rocket	N1
Spacecraft weight	197,975 pounds
Planned mission objective	Lunar orbital flight with complete test of the 7K-LOK spacecraft. Fourth test flight by the N1 booster rocket
Mission results	Rocket reached a flight duration of 110 seconds, the longest of the N1 program. Then it exploded at an altitude of 31 miles. The rescue system delivered the capsule to safety

The flight plan for the N1 rocket with the serial no. 7L was signed by Vasily Mishkin and his three deputies, Chertok, Okapin, and Tregub, on July 18, 1972. It envisaged a lunar orbital mission by the LOK spacecraft. Production unit 6A was selected for the mission. A functional lunar lander was not used, however. Instead, an LK mockup was carried beneath the L3 payload fairing. The total weight of the payload, which was supposed to be transported into a low Earth orbit, was 197,981 pounds. It consisted of Block G, Block D, the LK mockup, and LOK spacecraft no. 6A.

The launch of rocket 7L took place on November 23, 1972, at 11:11:52 local Kazakh time. The ignition of the engines, the liftoff from the launchpad, and the initial phase of flight proceeded successfully. The breaking of the sound barrier, passage through the zone of maximum dynamic pressure, and finally, ninety seconds after leaving the launchpad, the shutting off of the engines in the inner ring in Block A went according to plan. Problems did not begin to develop until the 104th second of flight, but things began happening very quickly. Within three seconds a fire broke out in the tail section, which immediately afterward developed into a powerful explosion. In seconds the entire rocket broke apart. It would have been only seven more seconds to first-stage burnout and ignition of the second-stage engines. The rescue system separated the return capsule from the L3 complex and returned it safely to Earth.

December 7: The US manned lunar program ends with the successful mission of Apollo 17. The mission is also the scientific high point of the Apollo program. Eugene Cernan, Ronald Evans, and Harrison Schmitt left Earth on a twelve-day mission to the Taurus Littrov mountains. Cernan and Schmitt landed and remained three days on the surface of the moon.

1973

At 07:55 Central European Time, a Proton K booster rocket sent the space probe Luna 21 with moon rover Lunokhod 2 on their way to Earth's satellite. The probe reached a transitional orbit with a perigee of 56 miles and an apogee of 62 miles on January 12.

On January 15, after forty orbits of the moon, the probe reduced the lowest point in its elliptical orbit to 10 miles and subsequently began its descent to land. The probe touched down on the surface of the moon at 00:35 Central European Time in the 34-mile-wide Le Monnier crater on the eastern edge of the Mare Serenitatis. The total "landed weight" on the moon was 4,000 pounds.

Mission Data, Luna 21–Lunokhod 2	
Mission designation	Luna 21
Date	January 8, 1973, 07:55 CET
Spacecraft	E-8EL no. 204 Lunokhod
Booster rocket	Proton K 8K78K
Spacecraft weight	12,566 pounds, rover 1,852 pounds
Planned mission objective	Lunar exploration by the Lunokhod rover
Mission results	Landed on January 15, 1971; last contact on May 9, 1973; distance covered: 26.7 miles
Last contact	15 degrees, 51 seconds north latitude; 30 degrees, 27 seconds east longitude near the Le Monnier crater

Lunokhod 2.

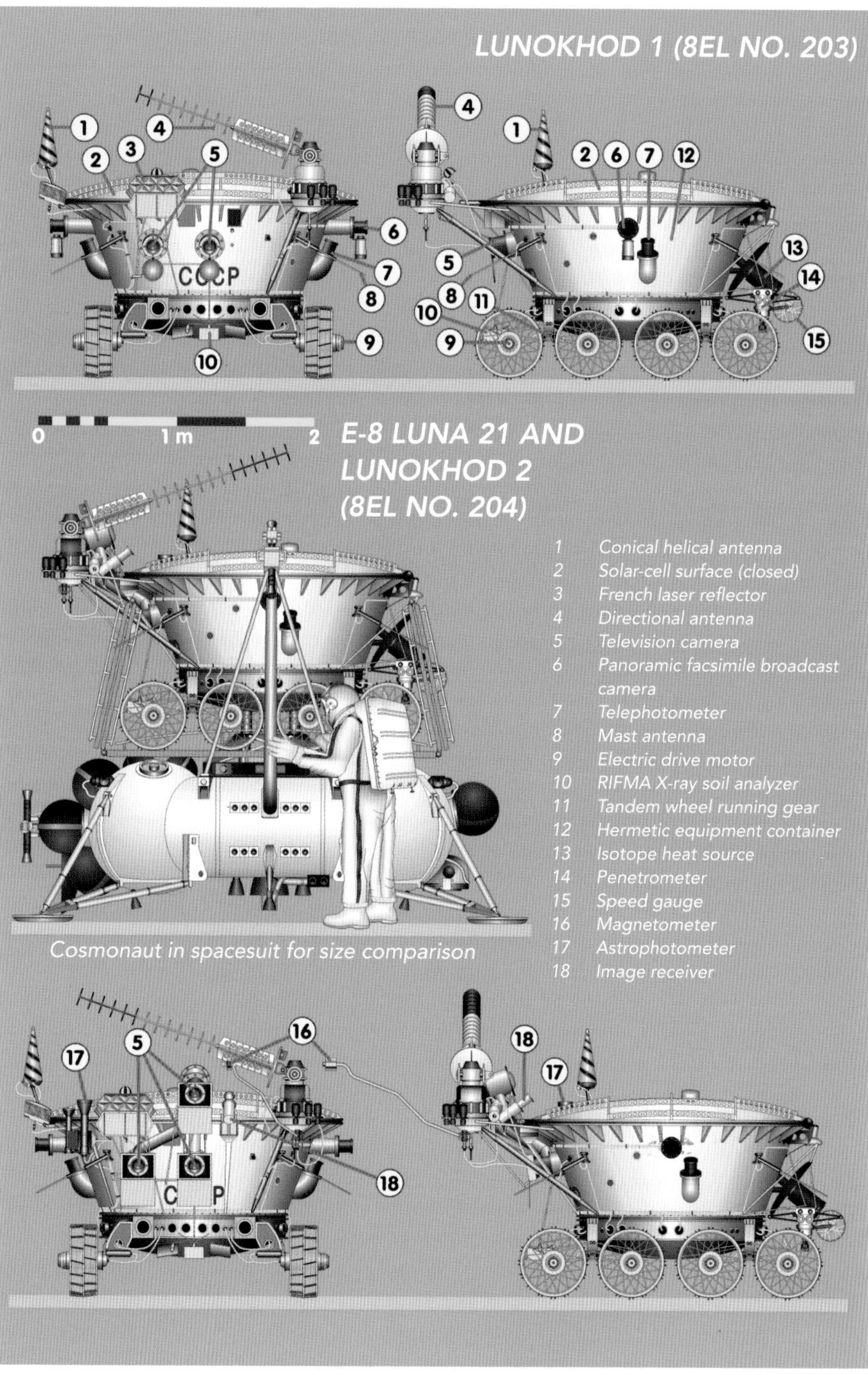

Cosmonaut in spacesuit for size comparison

Lunokhod 2.

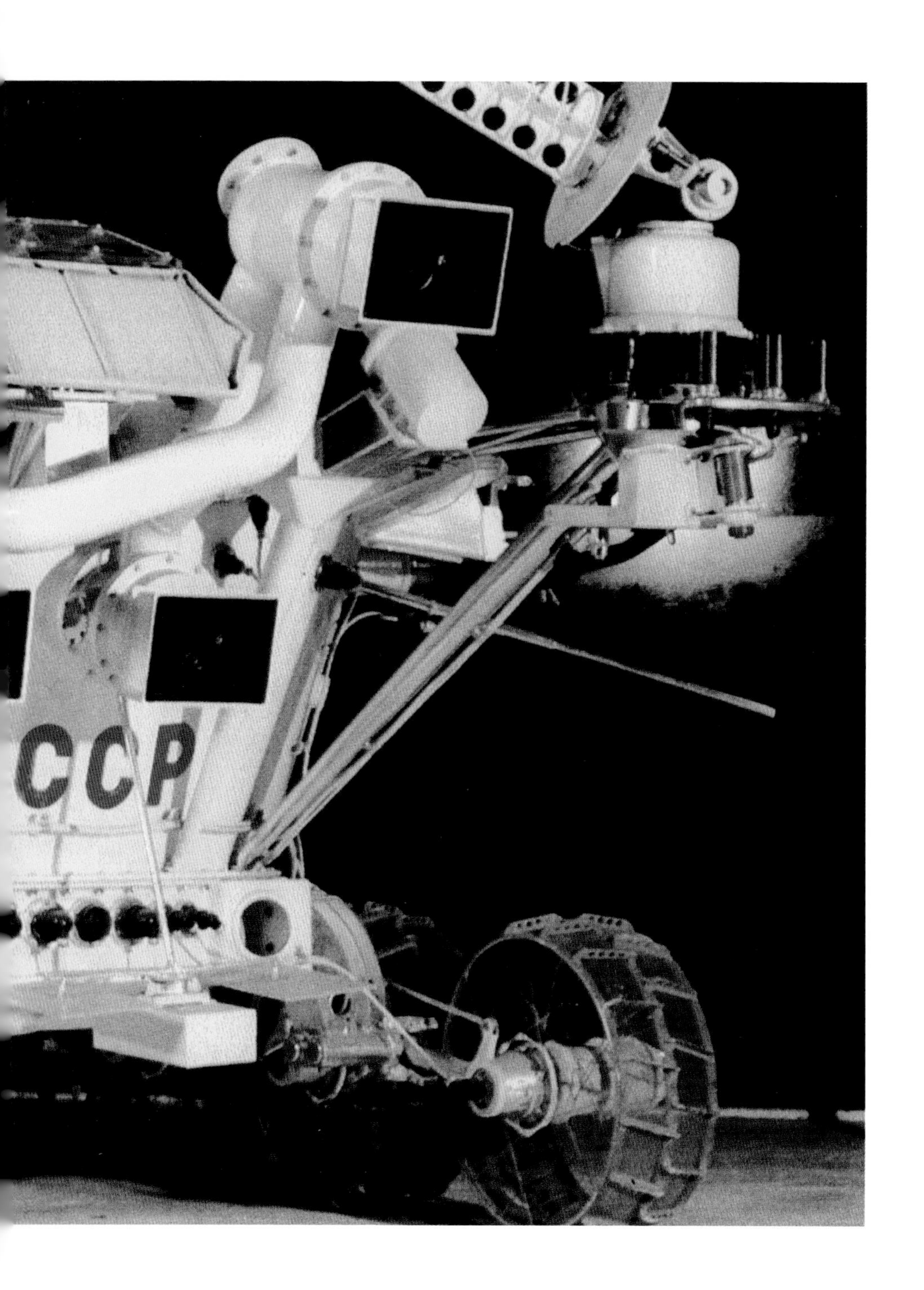
CCP

After the landing, Lunokhod 2 took the first pictures while still on the back of the probe. Then at 02:14 it rolled down onto the surface of the moon. Lunokhod 2 was active for three months, sending eighty-six panoramic photographs and 80,000 individual pictures back to Earth.

Lunokhod 2 was a much-improved variant of the Lunokhod 1 rover. It weighed 1,850 pounds, 187 pounds more than its predecessor. This extra weight was due to an improved power transfer system to the wheels, an additional TV camera, and additional scientific instruments.

There is another interesting anecdote about Lunokhod 2; in the financially strapped days after the Soviet Union, Russia remembered the long-forgotten old timer on the moon and came up with the idea of selling it to the highest bidder. And so in December 1993, the landing probe and the lunar automobile were put up for auction at Sotheby's in New York. The highest bidder was a certain Richard Garriot, and so a total of 2.2 tons of the best—if somewhat inconveniently located—Soviet space technology became his property for $68,500. Incidentally, Richard Garriot is the son of Skylab and space shuttle astronaut Owen Garriot. He is a game developer, multimillionaire, and private astronaut, who "bought" a flight to the International Space Station from the Russian spaceflight organization Roskosmos in October 2009. At the time of this writing, however, he has not been able to personally examine his property on the moon.

1974

On May 29, the Soviet Union launched lunar orbiter Luna 22, which entered lunar orbit on June 2, 1974. It was the second and also the last of the advanced orbital probes based on the E-8 platform. The mission was very successful and went on until November 1975. The spacecraft was moved a great deal during this mission, proof of the certainty that the controllers had achieved with the E-8 probes. A whole series of orbital change maneuvers were made to carry out various scientific experiments. The probe's orbital altitude was lowered to 15.5 miles above the surface of the moon. Luna 2 was the last lunar orbiter that the Soviet Union and its successor state Russia ever placed in orbit around the moon.

Mission Data, Luna 22	
Mission designation	Luna 22
Date	May 29, 1974, 09:57 CET
Spacecraft	E-8LS no. 220
Booster rocket	Proton K 8K78K
Spacecraft weight	12,566 pounds
Planned mission objective	Lunar orbiter
Mission results	Insertion into lunar orbit on June 2, 1974. Orbit 136 x 138 miles, inclination 19.35 degrees. Mission end on November 1975
Last contact	Crashed onto the surface of the moon, date unknown
Current location	Not known

On May 19, Defense Minister Grechko signed a decree suspending further launches of the N1. On June 24, 1974, Valentin Glushko, who had been made head of Energia, the successor organization to the legendary OKB-1, just a few days earlier, ended all work on the N1-L3 project. All programs that had to do with further improvements to and development work on the N1, such as the development of an upper stage powered by liquid oxygen and liquid hydrogen, were also canceled. This also affected the future improved L3M moon-landing system and many other programs.

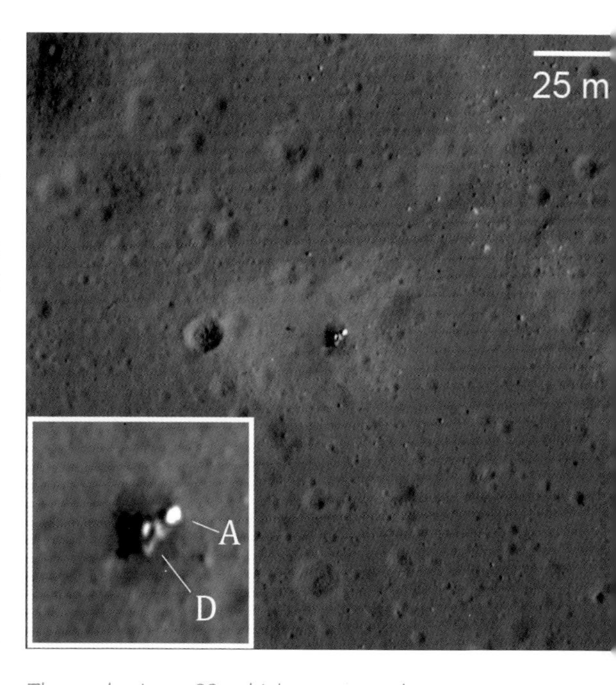

The probe Luna 23, which overturned on landing, was photographed more than forty years after its accident by the American Lunar Reconnaissance Orbiter.

After the program ended, the two almost complete units, the 8L and 9L, were disassembled and partially scrapped, as were the components of flight units 10L, 11L, 12L, 13L, and 14L, which were in the advanced-production stage. The greater part, however, was simply deposited thoughtlessly and dishonorably somewhere in the countryside around Baikonur. Many years later, visitors to the area around Baikonur could still see shelters and supply sheds made from tank sections of the N1.

All documentation concerning the N1 was also destroyed to prevent its possible "resurrection" in the Soviet space program at a later date. Glushko made certain that no references to the N1 program remained. Every photograph and every memento that remained in the OKB-1 museum were destroyed. With this, the Soviet manned lunar program was irretrievably dead. The Soviet unmanned lunar program lived for another two years, during which time the hardware built in earlier years was gradually used up.

June 10, 1973: The US sent another unmanned space probe into lunar orbit (Explorer 49). It would be almost twenty-one years before another spacecraft, the photo orbiter Clementine, was placed in orbit about Earth's natural satellite.

Luna 23 took off in the direction of the moon on October 24, 1974. This space vehicle was an improved sample return probe, which was supposed to be capable of taking core samples from a depth of 7.5 feet. The transfer to the moon with a course correction on October 31, the turn into lunar orbit on November 2,

Mission Data, Luna 23	
Mission designation	Luna 23
Date	October 28, 1974, 15:30 CET
Spacecraft	E-8-5 no. 410
Booster rocket	Proton K 8K78K
Spacecraft weight	12,786 pounds, lander 4,145 pounds
Planned mission objective	Automatic sample return
Mission results	Crashed for reasons unknown during landing. Transmitted data to Earth for three days. Return flight by the sample capsule was not possible
Position	12.67 degrees north latitude, 62.15 degrees east longitude

and almost the complete landing phase on November 6 went according to plan.

Then, however, the probe experienced a bit of very bad luck; it was damaged on setting down, and as a result it was unable to take core samples, nor could the ascent stage return to Earth. Otherwise, however, it functioned and sent data to Earth for three days after landing. Years later, Luna 23 was discovered by the American moon probe Lunar Reconnaissance Orbiter. It showed that Luna 23 had tipped over during landing.

1975

On October 16, 1975, another mission failed during launch. The Proton K failed for one last time in the Soviet lunar program. This time it was a turbopump in the Block D stage that prevented that unit from igniting. As a result, the probe failed even to achieve Earth orbit and burned up over the Pacific. Its target had been the Mare Crisium, which Luna 23 had unsuccessfully tried to reach.

Mission Data, Luna E-8-5 M No. 412	
Mission designation	—
Date	October 16, 1975, 05:05 CET
Spacecraft	E-8-5M no. 412
Booster rocket	Proton K 8K78K
Spacecraft weight	12,345 pounds
Planned mission objective	Automatic sample return
Mission results	The probe failed to reach orbit due to the failure of a turbopump in the Block D upper stage
Fate	Crashed into the Pacific about thirty minutes after launch

1976

In February 1976, the N1-L3 program was officially canceled. The Soviet manned lunar program had finally ended.

On August 9, 1976, Luna 24 was launched to the moon, and on August 18 it successfully landed in the Mare Crisium, just a few hundred feet from the spot where Luna 22 lay overturned on the lunar surface. The probe collected a soil sample weighing 6 ounces, more than that brought back by Luna 16 and Luna 20 combined. The launch back to Earth began on August 19. After an uneventful return flight, the probe landed 120 miles southeast of the city of Surgut in western Siberia at 06:55 Central European Time on August 22.

Mockup of Luna 24.

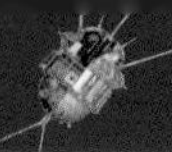

Mission Data, Luna 24	
Mission designation	Luna 24
Date	August 9, 1976, 16:04 CET
Spacecraft	E-8-5 M no. 414
Booster rocket	Proton K 8K78K
Spacecraft weight	12,786 pounds, lander 4,145 pounds
Planned mission objective	Automatic sample return
Mission results	Landed on August 18, 1976; return launch for Earth one day later. Transported six ounces of lunar samples to Earth. Landed near the Luna 23 accident site
Current location	12.71 degrees north latitude, 62.21 degrees east longitude

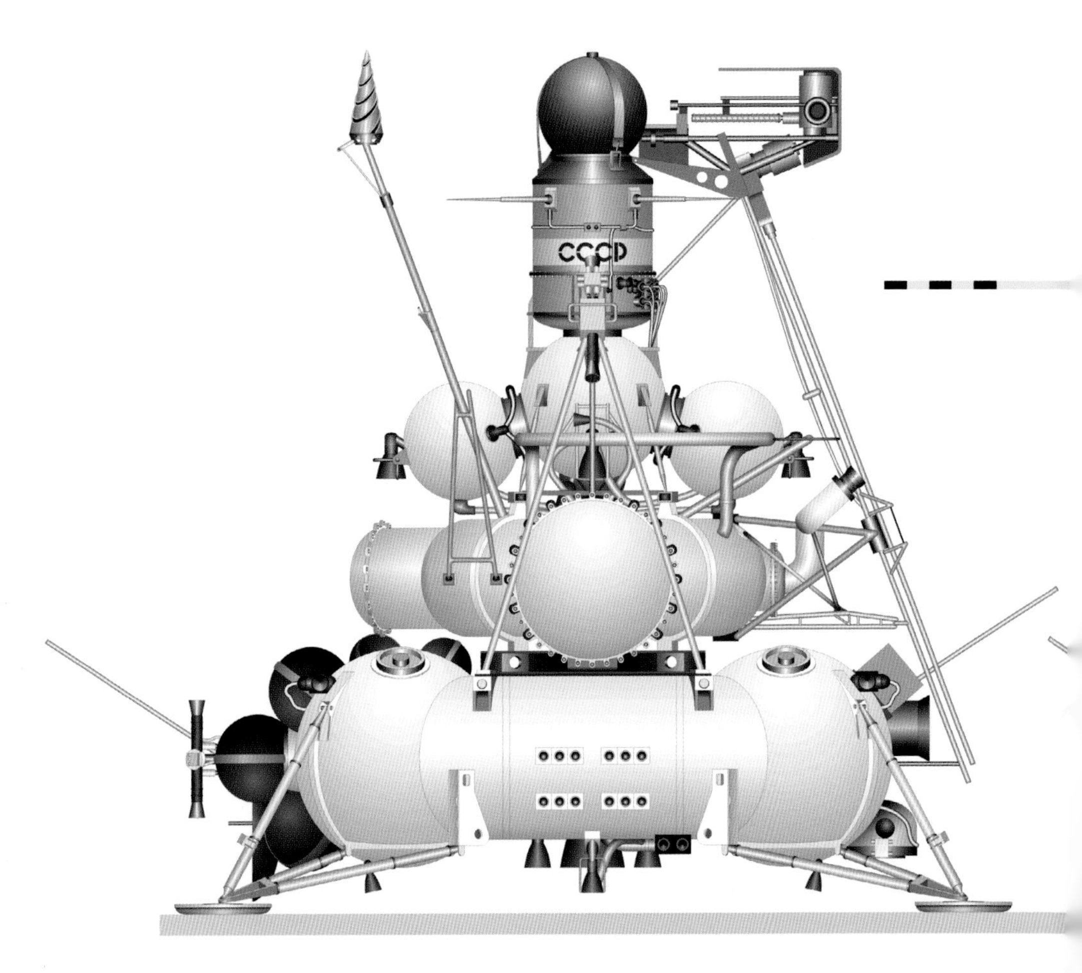

Mockup of Luna 24.

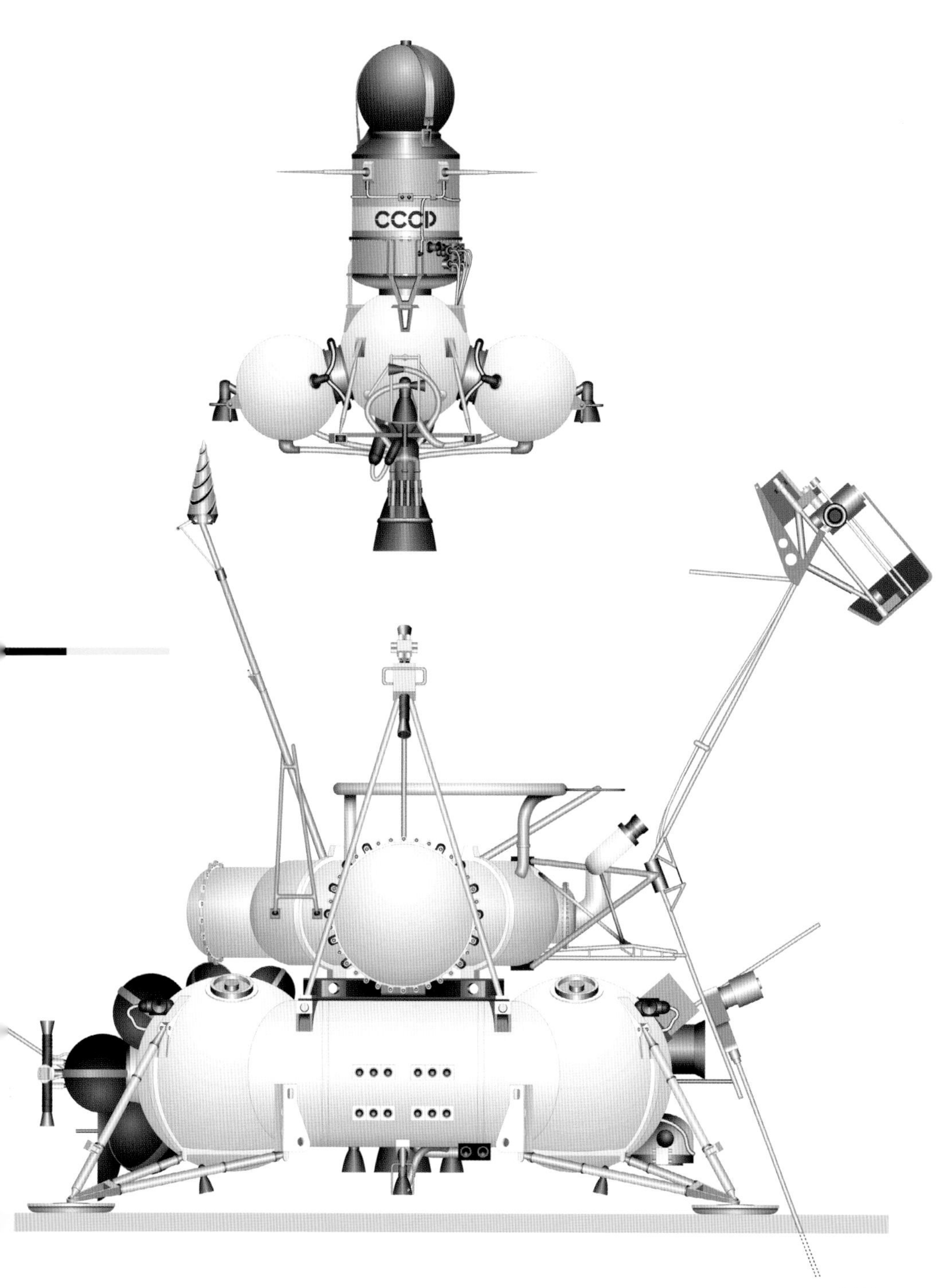

The manned Soviet moon landing remained a dream. Artist's concept of an LK lander on the moon.

EPILOGUE

On August 24, 1976, the Soviet news agency TASS reported that there was still contact with Luna 24's landing stage on the moon. This report was the last official statement ever made during the Soviet Luna program.

Over forty-one years have passed since these sentences were written. During that time neither the Soviet Union nor its successor state Russia has sent a single space probe to the moon. Even more remarkable is that Russia would not be capable of repeating the feat of the thirteen-day mission of Luna 24 in 1976, even within five years.

It was thirty-seven years and four months after the soft landing on the moon by Luna 24 that the next such event took place. It was carried out by the Chinese space probe Chang'e 3 in December 2013.